KEY DATES IN NUMBER THEORY HISTORY

A superb illustration of the influence of mathematics on the great German artist Albrecht Durer's (1471-1528) work is Melencolia I (Melancholy). He produced this masterpiece following the death of his mother in 1514. In it, he portrays myriad mystic symbols and documents his skill in perspective drawing and his knowledge of mathematics. Melencolia I depicts a 4 by 4 magic square. The date of the work (1514) is indicated in the two middle boxes of the bottom row.

KEY DATES IN NUMBER THEORY HISTORY

From 10,529 B.C. TO THE PRESENT

DONALD D. SPENCER, Ph.D.

CAMELOT PUBLISHING COMPANY
Ormond Beach, Florida

CAMELOT PUBLISHING COMPANY
P.O. Box 1357
Ormond Beach, FL 32175

Printed on acid-free paper

ISBN 0-89218-318-7

Library of Congress Cataloging-in-Publication Data

Spencer, Donald D.
        Key dates in number theory history : from 10,529 B.C. to the
    present / Donald D. Spencer.
            p.   cm.
        Includes index.
        ISBN 0-89218-318-7 (acid-free)
        1. Number theory--History--Chronology.   I. Title.
    QA241.S6343 1995
    512'.7' 09--dc20   95-77
                                                                 CIP

INTRODUCTION

Number Theory is a branch of mathematics concerned with the properties of integers, or whole numbers, such as 0, 1, 2, 3, ... These properties have been the object of fascination and investigation for thousands of years; interest in the natural numbers is as old as civilization itself.

It is perhaps the only branch of mathematics where there is any possibility that new and valuable discoveries might be made without an extensive acquaintance with technical mathematics. In recent years, high school students have made new and important discoveries with perfect numbers and prime numbers. College students are regularly making important contributions to the number theory field. Personal computer users around the world are using their machines to produce new number theory results on a regular basis.

Key Dates in Number Theory History is an illustrated chronology of the most important number theory events from 10,529 B.C. to the present. This book tells what happened in number theory, when it happened, and who made it happen.

The book is a work that allows us to comprehend the events that have taken place in number theory. In a manner that is both fascinating and accessible, it charts the progress to date and documents the remarkable achievements of the men and women of mathematics, computer science and number theory.

Illustrations and photographs are used throughout the book to highlight specific number theory events and to portray some of the people who made important contributions to number theory.

The comprehensive index will allow the book to be used for general and particular reference. It is hoped that **Key Dates in Number Theory History** will prove a useful guide through the many centuries of number theory historical development.

The book is an indespensable reference work for everyone interested in the history of number theory.

The IBM 704 computer was first delivered to customers in 1955. It was a successor to the IBM 701 computer. The IBM 704 incorporated into hardware floating point arithmetic and three addressable index registers (which simplified programming). The FORTRAN programming language was developed for use on the IBM 704 — an innovation that led to a huge expansion in the uses and users of computers. In 1959, Francois Genuys programmed an IBM 704 computer to generate π to 16,167 decimal digits in 4.3 hours.

CONTENTS

The Stretch computer (IBM 7030) was designed to have 100 times the performance of the IBM 704 computer, which was the most powerful computer available in 1955. The first Stretch computer was delivered in 1961. It was the most powerful computer in existence. Its 150,000 transistors could execute 100 billion instructions per day. It could cope with more than one instruction at a time and prepare itself for future work. Stretch was the first major solid state computer developed by the IBM Corporation, and its transistor, core, and disk storage technologies were applied extensively to other computers of the 7000 series. In 1966, French mathematicians Jean Gilloud and J. Fillatoire used a Stretch computer at the Commissariat a l'Energie Atomique in Paris to compute π to 250,000 decimal places.

KEY DATES IN NUMBER THEORY HISTORY

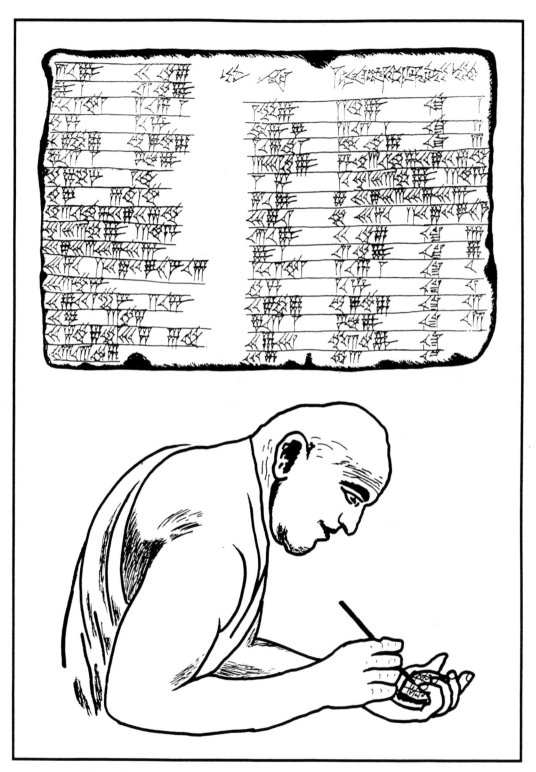

The Babylonians wrote their numbers on clay tablets

10,529 B.C. A Babylonian tablet, which is rich in mathematical material, gives the reciprocals of all numbers between 58 and 80 with great accuracy. The tablet is now part of the Yale Babylonian Collection.

8500 B.C. A prehistoric Neanderthal people called Ishango inhabited much of Europe and areas surrounding the Mediterranean. Among their remains is a bone with three notched columns. It is agreed that the notches represent a first use of a unary number system.

3000 B.C. First recorded evidence of geometry can be found with the Babylonians. They measured circles and observed the stars as well. They also had a system for recognizing many of the planets and their orbits.

2900 B.C. By now the Egyptians (north of the Sahara) had developed an advanced civilization. They gave us our first solar calendar, a number system, and an introduction to fractions. The Egyptians designed temples and pyramids with the use of plumb lines and triangles.

2200 B.C. The lo-shu is the oldest known example of a magic square, and myth claims that it was first seen by the Emperor Yu decorating the back of a divine tortoise along a bank of the Yellow River.

2000 B.C. Babylonian arithmetic had evolved into a well-developed prose algebra.

2000 B.C. Babylonians use $\pi = 3.125$ and $(16.9)^2 = 3.1605$.

1650 B.C. A scribe named Ahmes wrote a document containing puzzles and their histories. Ahmes, however tells us he is only copying a much older papyrus, which was written in the time of Nema'et-Re, a ruler of Egypt, which takes the document back to almost 1850 B.C., and perhaps even earlier if that text was itself a copy of an older one. The Ahmes Papyrus contains 84 problems and their solutions. The papyrus was purchased by A. Henry Rhind in 1858 and is often called the "Rhind Papyrus."

1400 B.C. Construction started, by Ramses I, of a great monument on the side of the Nile River near the city of Thebes. The slabs on the

Pythagoras was a teacher of arithmetic and number theory. Pythagoras and his followers studied numbers as geometric arrangements of points, such as the triangular numbers 1, 3, 6, 10, 15, 21, 28, 36, 45, and so on.

Euclid has been called "The Father of Geometry." He founded and taught the school of mathematics in Alexandria and, while there, wrote several books. Euclid's writings, in accordance with the practice of the time, were written on rolls of parchment or papyrus. Although Euclid may not be called the greatest geometrician, he certainly was one of the greatest organizers and compilers of materials on that subject. There are thirteen books in his *Elements*, the earliest textbook on geometry. His first book deals with plane geometry. The others concern themselves with ratio and proportion, solid geometry, polygons, circles and number theory.

roof of the temple have a few curious sketches of mathematical figures and game boards.

1350 B.C.   The Rollin papyrus contained some elaborate bread accounts showing the practical use of large numbers at this time.

1300 B.C.   Chinese uses $\pi = 3$.

550 B.C.   I Kings vii, 23 implies $\pi = 3$.

500 B.C.   The Hindu *Sulvasutras*, religious writings, include an indication an indication that "Pythagorean triples" such as (15, 36, 39) were used in constructing temples between 800 B.C. and 500 B.C. There is no evidence that a proven general formula was used to obtain these triples.

540 B.C.   About this time Pythagoras may have given the first general proof of the Pythagorean theorem. Pythagoras and his followers took the first steps in the development of elementary number theory. They defined amicable numbers, perfect numbers, figurate numbers and some of their properties.

434 B.C.   Anaxagoras attempts to square the circle.

425 B.C.   Theodorus of Cyrene showed that the square root of numbers from 3 to 17 (other than perfect squares) were irrational.

335 B.C.   Dinostratos uses the quadratrix to square the circle.

300 B.C.   Euclid produced the *Elements*. Probably no single work has exerted a greater influence on scientific thinking. The *Elements* contain geometry and number theory algorithms. The number theory algorithms include the Euclidean algorithm for finding the greatest common divisor of two numbers, the fundamental theorem of arithmetic and the proof of the infinitude of prime numbers. The *Elements* consisted of thirteen volumes. Today, high-school plane and solid geometry texts contain much of the material found in these volumes.

300 B.C.   The Cairo Mathematical papyrus was unearthed in 1938 and examined in 1962. This papyrus contains forty mathematical problems, nine of which deal exclusively with the Pythagorean theorem and show that the Egyptians of that time not only

Archimedes (287-212 B.C.) was a native of Syracuse, a Greek city on the island of Sicily, who made many important contributions to mathematics. One of the first times very large numbers were used was by Archimedes in about 250 B.C., when he reportedly computed the number of grains of sand in the universe to be 10^{63}. Archimedes was one of the greatest mathematicians of all times.

Eratosthenes (275-194 B.C.), a Greek geographer and mathematician, was one of the greatest scholars of Alexandria. He was a highly talented and versatile person. In addition to his Sieve for prime numbers, he made some remarkably accurate measurements of the size of the earth.

knew that the 3, 4, 5 triangle is right angled, but that the 5, 12, 13 and 20, 21, 29 triangles were right angled as well.

250 B.C.   Archimedes, considered the greatest mathematician of antiquity showed that π was between 223/71 and 22/7.

230 B.C.   Greek mathematician Eratosthenes invented his "sieve" for finding all primes less than a given number.

225 B.C.   Appolonius improves Archimedes value of π, unknown to what extent.

200 B.C.   Chinese mathematician Chang Tshang is believed to be the author of *Nine Chapters on the Mathematical Art*, a work that summarized a great deal of Chinese mathematics of the time: the decimal number system, the use of zero and negative numbers, and the right triangle theorem.

100 B.C.   Nichomachus extended the number theory of the Pythagoreans. For example he discovered that given the odd numbers 1, 3, 5, 7, 9, 11, 13, 15, 17, ..., the first is the cube of 1, the sum of the next two is the cube of two, the sum of the next three is the cube of 3, and so on.

130 B.C.   Hou Han Shu used $\pi = 3.1622$.

100   Greek mathematician Nicomachus of Gerasa gave a complete description of triangular and square numbers in *Introduction arithmetica*, the earliest extant manuscript of which dates back to the tenth century. It is generally accepted that there is little in this work which is original with Nicomachus himself; but his work brought together the results of previous generations in a reasonable clear and concise manner.

150   The first notable value for π, after that of Archimedes, was given by Claudius Ptolemy of Alexandria in his famous *Syntaxis Mathematica*, the greatest ancient Greek work on astronomy. In this work, π is given as 377/120 or 3.1416. Undoubtedly, this value was derived from the table of chords, which appears in the treatise.

250   Diophantus of Alexandria solved many special equations having integral solutions. The known titles of Diophantus are the *Arithmetica* in 13 books, the *Porisms*, and a treatise on polygonal numbers. All of

Tse Ch'ung-chih (about 480 A.D.), the Chinese mathematician gave 22/7 as an "inaccurate value" of π, and 355/113 as the "accurate value," and he showed that π lies between our present decimal forms 3.1415926 and 3.1415927.

In India, the Hindus used matheamtics and passed their knowledge of numerals to the Arabs.

them deal with the properties of rational or integral numbers. Note: the Greek mathematicians use the term *arithmetic* in the sense of number. A Diophantine equation is a polynomial equation with integer coefficients for which only integer solutions are allowed.

250   Chinese mathematician, Sun-tsi, wrote a book containing the problem: Find the smallest number which when divided by 3 leaves 2, when divided by 5 leaves 3 and by 7 leaves 2. This is the first instance of the Chinese Remainder Theorem.

250   Diophantus' *Arithmetica* contained many problems in elementary number theory.

250   A major step in the development of algebra occurred with the work of Greek mathematician Diophantus. He worked out a system of his own to solve problems by using symbols to replace numbers and operations.

263   Liu Hui uses $\pi = 157/50 = 3.14$.

264   Liu Hui used a variation of the Archimedean inscribed polygon; using a polygon of 192 sides, he found $3.141024 < \pi < 3.142704$ and with a polygon of 3,072 sides, he found $\pi = 3.14159$.

320   Iamblichus of Chalcis, an influential Neoplatonic philosopher ascribes to Pythagoras the discovery of amicable, or friendly, numbers. Two numbers are amicable if each is the sum of the proper divisors of the other.

380   Hindu mathematicians used the value $\pi = 3 \ 177/1250 = 3.1416$.

480   The Chinese mathematician Tsu Ch'ung-chih calculated π as $355/113$ which is correct to six places $(3.1415929...)$. This approximation was not rediscovered in Europe until 1585.

499   Hindu mathematician Aryabhata gave $62,832/20,000 = 3.1416$ as an approximate value for π. It is not known how this result was obtained.

500   Hindu mathematicians published indeterminate problems.

540   Ancient Greek mathematicians were entranced by figurate numbers: numbers that could be represented by arranging points in regular patterns on a plane or in space.

Muhammed idn Musa al-Khowarizmi (780-850) was a Persian mathematician and astronomer who lived in Baghdad and whose name has given rise to the word **algorithm**. He wrote a book *Hisab al-jabr wal-mugabalah*, which was a compilation of rules for solving linear and quadratic equations, and has given rise to the word **algebra**.

Leonardo Fibonacci (1170-1250) was a famous 12th and early 13th century Italian mathematician. He was generally regarded as the greatest and most productive mathematician of the Middle Ages. In 1202, he wrote a book called *Liber Abaci* which introduced the Hindu-Arabic numerals into Europe. Fibonacci's writings dealt with a wide range of topics, including abstract mathematics, algebra, geometry, and problem solving.

554 Korean scholars introduce Chinese mathematics into Japan.

575 Ch'ang Kiu-Kien finds $\pi = 3.14$.

598 Hindu mathematician Brahmagupta in his mathematical and astronomical manual *Brahma-Sphuta-Siddhanta* (Brahma's correct system) introduces and describes "linear indeterminate problems." These problems occur quite commonly in puzzles and whose theory constitutes a particularly significant part of number theory. He also used the value $\pi = \sqrt{10} = 3.162277...$

718 A document contains $\pi = 92/29 = 3.1724...$

820 Mohammed ibn Musa al-Khowarizmi wrote influential treatise on algebra and a book on the Hindu numerals. He improved algebraic knowledge and solved equations by using rules.

870 Tabit ibn Qorra (translator of Greek works, algebra, magic squares and amicable numbers).

900 Egyptian mathematician Abu Kamil published *The Book of Precious Things in the Art of Reckoning* which contained an early example of a general method used to solve "linear indeterminate equations."

900 By now, black Africans had helped establish mathematical rates of exchange through gold, salt, and kola nut trade with Arabs across desert caravan routes.

1150 Bhaskara, a famous Hindu astrologer, was also a mathematician of exceptional ability. One of his books, *Lilivaki*, gives rules for all mathematical functions. Bhaskara gave several approximations to π: 3927/1250 for accurate work, 22/7 for a rough value, and the square root of 10 for everyday usage.

1200 Leonardo Fibonacci finds $\pi = 3.141818$.

1202 Leonardo Fibonacci wrote a book called *Liber Abaci* which introduced the Hindu-Arabic numerals into Europe. The now famous Fibonacci number sequence 1, 1, 2, 3, 5, 8, 13, ..., where each number is the sum of the previous two, were described in this book in a problem involving the breeding of rabbits. The *Liber Abacci* was a hand-written manuscript, copied later by scribes who devoted their lives to

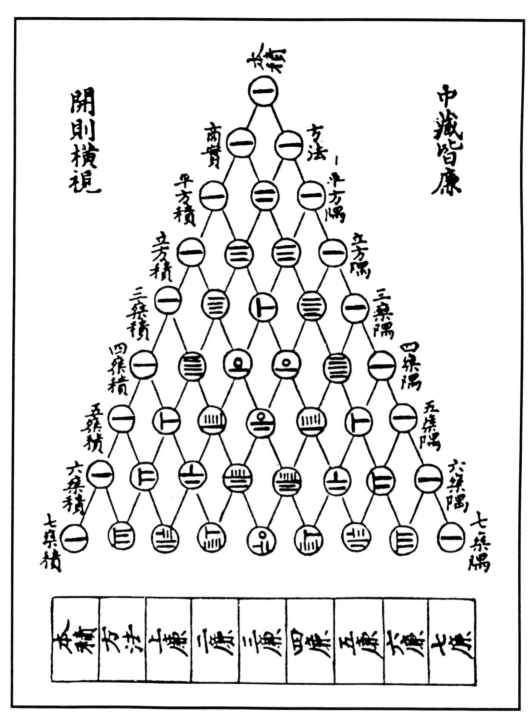

Page from *The Precious Mirror of the Four Elements*, which contains a triangle of numbers which later became known as Pascal's Triangle. Written by Chu Shih-chieh in 1303, more than 320 years before Blaise Pascal was born.

such tasks. Fibonacci is credited by historians as having an enormous influence in the resulting changeover from Roman to Hindu-Arabic numerals.

1206 Yang Hui produced the earliest known version of what we call Pascal's triangle.

1222 Jordanus Nemorarius wrote several works dealing with arithmetic, algebra, geometry, and statistics. His algebra was the first one widely to use letters to represent general numbers.

1228 Leonardo Fibonacci revised *Liber Abaci* and wrote other books. His writings dealt with a wide range of topics, including abstract mathematics, algebra, geometry and problem solving. He designed one of the earliest square root symbols (no longer in use today).

1240 John of Holywood wrote the standard mathematics text that was used for centuries in European universities.

1247 Ch'in Kiu-shao showed the high degree of sophistication of Chinese mathematics in his works on solving higher degree equations by numerical methods, a discovery not made in Europe until 1819.

1260 Yang Hui used decimal fractions and produced the earliest known version of what we call Pascal's triangle.

1299 Nearly a century after Leonardo Fibonacci published his book, *Liber Abaci*, describing the Hindu-Arabic numerals, Florence, Italy strictly forbade their use in commercial and legal transactions.

1303 Chu Shih-chieh wrote *The Precious Mirror of the Four Elements* which contained a triangle of numbers, which later became known as Pascal's Triangle. Chu Shih-chieh displayed the binomial coefficients more than 200 years before it was published in Europe, and 320 years before Blaise Pascal was born.

1360 Nicole Oresme introduced fractional exponents in his unpublished work *Algorismus Proportionum*.

1429 Al-Kashi, astronomer royal to Ulugh Beg of Samarkand, computed π to sixteen decimal places.

Woodcut from *Margarita Philosophica*, by Gregorius Reich (1503). The drawing shows one man (Pythagoras) computing with a form of abacus, the other (Boetius) with Hindu-Arabic numerals. Arithmetic, symbolized by the woman standing in the center, seems to be deciding the debate between the two men.

1440   Hindu-Arabic numerals spread rapidly throughout Europe during the Renaissance, especially after the invention of moveable type for printing, which led to new textbooks.

1450   Medieval scholar and philosopher Nicolaus Cusanus discovered a method to compute π.

1478   The earliest printed arithmetic, *Treviso Arithmetic*, was published in the town of Treviso, Italy, which was located on the trade route linking Venice with the north. It contained some recreational mathematics.

1479   Hindu-Arabic digits first appeared in their present form.

1480   Approximate date when the plus sign, +, and minus sign, −, was introduced in Germany.

1482   First printed edition of Euclid's *Elements*. Over one thousand editions of *Elements* have appeared since the first one was printed; for more than two millennia, this work has dominated all teaching of geometry.

1484   Nicolas Chuquet introduced the notation for adding and subtracting fractions.

1489   Johann Widmann used the + and − signs to denote addition and subtraction.

1491   Filippo Calandri introduced the long division process that is still in use today.

1500   Through the Arabs the knowledge of amicable numbers spread to Western Europe. They are mentioned in the works of many prominent mathematical writers: Nicolas Chuquet, Etienne de la Roche (Villefranche). Michael Stifel, Cardanus and Tartaglia.

1508   A popular woodcut showed the debate between the "abacists" (those who kept to the old method) and the "algorists" (those who followed Leonardo Fibonacci).

1514   German artist/mathematician Albrecht Durer produced the engraving *Melencolia* which contained an order 4 magic square. The year of the engraving is in the middle cells of the bottom row. The *Melencolia* square is one of the first magic squares to be printed.

Adam Riese (1489-1559) was one of the most influential of the German writers in the movement to replace the old computation by means of counters by the more modern written computation. He wrote one of the most popular arithmetic school-books of the 16th century.

SUMARIO COMPENDIOSO

The *Sumario Compendioso*, printed in Mexico in 1556, is a book of computational tables that aided merchants engaged in the buying of gold and silver that was being taken from the mines of Peru and Mexico. The most interesting feature of the work, however, is neither the tables nor the arithmetic; it consists of six pages devoted to algebra, chiefly relating to the quadratic equation.

1522   Adam Riese published a table of square roots.

1525   Christoff Rudolff introduced the square root symbol that is used today in his book on Algebra entitled *Die Coss*. This book was very influential in Germany, and an improved edition of the book was brought out by Michael Stifel in 1553. Stifel has been described as the greatest German algebraist of the sixteenth century.

1526   Gerolamo Cardano wrote *Liber de ludo aleae* (The book of games of chance), however, it was not published until 1663. Cardano, of cubic equation fame, was not only a mathematician, engineer and physician, but also a passionate gambler.

1527   Pascal Triangle appears on the title page of an arithmetic work by Petrus Apianus.

1527   Petrus Apianus introduced notation for multiplying and dividing fractions.

1527   The coefficients for binomial expansions were published.

1530   Christoff Rudolff published a collection of problems that involved computations with decimal fractions. He used a decimal notation similar to the modern one, with a bar to separate the integral and fractional parts.

1542   Mathematician Robert Recorde published *The Grounde of Arts, Teachyng the Works and Practice of Arithmetike*, a text that attempts to set out certain rules for arithmetic.

1545   German mathematician Michael Stifel used the letters M to denote multiplication and D for division, e.g., $2xy$ would be expressed as $2MxMy$.

1556   Tartaglia first used parentheses, ( ), for grouping terms.

1556   First work on mathematics printed in the New World — *Sumario Compendioso* by Spaniard Juan Diez. The book which was printed in Mexico, contained computational tables and six pages devoted to algebra.

512.7
C-1

Mid-nineteenth century ammunition store in Calcutta. Stacking canon balls in pyramids began before 1600. Knowledge of pyramidal numbers made it easy to work out how many balls in a pyramid of known height.

1557 The sign = was introduced in England by Robert Recorde. By the middle of the seventeenth century, Recorde's symbol was adopted by several influential writers and was in common use in England.

1559 Johannes Buteo, a French scholar, published a book *De quadratura circuli*, which is probably the first book that amounts to a history of π and related problems.

1572 Rafael Bombelli published the first consistent treatment of imaginary numbers.

1573 Valentinus Otho found the early Chinese value of π, namely 355/113 or 3.1415929.

1575 It was observed that every even perfect number is a triangular number.

1575 Diophantus' *Arithmetica* was one of the last Greek mathematical works to be translated into Latin. The *Arithmetica* considered many problems in elementary number theory. The Latin edition was published in Heidelberg by the German professor Holzman, a name which he later changed to the Greek form Xylander.

1579 French mathematician Francois Viete found π correct to nine decimal places, using polygons having 393,216 sides. He also found an infinite product for $2/\pi$ that used square roots and published a book in which he used decimal fractions as a matter of course, and recommended their use to others.

1582 October 4 was the last date reckoned by the old Julian calendar in Catholic countries.

1582 The Gregorian calendar which we use today was promulgated by Pope Gregory XIII.

1583 Simon Duchesne finds $\pi = (39/22)^2 = 3.14256...$

1583 Christopher Clavius initiated the use of a dot for multiplication.

1585 Dutchman Simon Stevin gave the first systematic presentation of the rules of operations on decimal fractions.

Francois Viete (1540-1603) was one of the greatest French mathematicians of the 16th century. He was a lawyer at the court of Henri IV of France and studied equations. Viete made important contributions to arithmetic, algebra, trigonometry, and geometry. He simplified the notation of algebra and was among the first to use letters to represent numbers. He also introduced a number of new words into mathematical terminology, some of which, such as *negative* and *coefficient* have survived. His attack on π resulted in the first analytical expression giving π as an infinite sequence of algebraic operations.

1585 Adriaen Anthoniszoon rediscovered the ancient Chinese ratio 355/113. This was apparently a lucky accident, since all he showed was that $377/120 > \pi > 333/106$. He then averaged the numerators and the denominators to obtain the "exact" value of π.

1585 Simon Stevin published a 7-page pamphlet in which he explained decimal fractions and their use.

1588 P.A. Cataldi discovers the 6th Mersenne prime, $2^{17} - 1$ and the 7th Mersenne prime, $2^{19} - 1$.

1591 French mathematician Francois Viete systematically used letters to represent unknowns. He also expressed the product of A and B as "A in B."

1593 Adriaen van Roomen of the Netherlands found π correct to fifteen decimal places by the classical method, using polygons having 2^{30} sides.

1596 Dutchman Ludolph van Ceulen, professor of mathematics at the University of Leyden, computed π to twenty decimal places.

1608 Bartholomaeus Pitiscus of Heidelburg was the first to use the decimal point.

1610 Ludolph van Ceulen of the Netherlands computed π to 32 and 35 decimal places using polygons having 2^{62} sides. He spent a large part of his life on this task, and his achievement was considered so extraordinary that his widow had the last three digits of the number engraved on his tombstone. In the Netherlands, π was also known as "Ludolph's number."

1611 German astronomer Johann Kepler arrived at the numbers illustrated by Leonardo Fibonacci's rabbit sequence in his own work. There is no indication that he had access to any of Fibonacci's writings.

1612 Claude-Gaspar Bachet, Sieur de Meziriac, a scholar devoted to classical learning, published a collection of ancient puzzles and, for the first time, rules for solving indeterminate problems. His work proved popular and a second enlarged edition appeared in 1624.

Title page of the 1619 edition of John Napier's *Mirifici logarithmorum canonis descriptio*, which also contains his *Constructio*. Napier wrote two works on logarithms, another work on computing rods (*Rodgologiae*), and another on algebra.

1614 John Napier, a Scottish politician, published the first book of logarithms. Napier coined the term "logarithm."

1617 John Napier devised a computing system based on logarithms. This system called "Napier's bones," was simply a set of numbering rods upon which he transcribed the results of multiplication. Napier's bones were described in his work *Rabdologiae*.

1617 John Napier introduced present day notation by using the decimal point when writing numerals.

1617 Henry Briggs, a professor at Oxford University, wanted to improve the usefulness of Napier's logarithms. Briggs and Napier agreed that a system of logarithms using powers of 10 with log 1 = 0 and log 10 = 1 would be the best for computations in the decimal system. Briggs published partial tables of these "common" logarithms. In 1624, Briggs published a more complete set of tables.

1619 John Napier's *Mirifici logarithmorum canonis constructio* was published posthmously by his son Robert. Rarely in the history of science has a new idea been received more enthusiastically. Universal praise was bestowed upon its inventor, and his logarithms was quickly adopted by scientists all across Europe and even in faraway China.

1621 It was probably the Latin translation of Diophantus' *Arithmetica*, made by Bachet de Meziriac, that first directed Pierre de Fermat's attention to number theory. Many of Fermat's contributions to the field occur as marginal statements made in his copy of Bachet's work.

1621 Dutch physicist Willebrord Snell calculated π to 35 decimal places by using polygons having only 2^{30} sides.

1624 Wilhelm Schickardt, a German professor of mathematics, designed a calculating machine that incorporated Napier's bones onto cylinders that could be rotated to perform calculations.

1624 Johann Kepler and Henry Briggs published a 14 place table of logarithms to the base ten. Briggs also coined the words "mantissa" and "characteristic."

1627 Vlacq published a table of logarithms from 1 to 100,000.

Marin Mersenne (1588-1648) was a Franciscan friar and spent most of his life in Parisian cloisters. His importance to the mathematical world of his day was due not so much to his own work, as to his energy, friendly enthusiasm, and extensive correspondence with other mathematicians. He was a voluminous writer, editing some of the works of Euclid, Apollonius, Archimedes, Theodosius, Menelaus, and various other Greek mathematicians. It is in the theory of numbers, however, particularly with respect to prime numbers and perfect numbers that he made contributions of real value.

1630   Grienberger computed π to 39 decimal places. This was the last major attempt to compute π by the method of perimeters.

1631   The problem of finding multiply perfect numbers was formulated by Marin Mersenne in a letter to Rene Descartes.

1631   Englishman William Oughtred used the x sign for multiplication.

1631   Thomas Harriot introduced the inequality symbols > and <. Harriot also introduced the dot symbol for multiplication, however, it was not widely used until Gottfried Leibniz adopted it.

1632   Albert Girard wrote that the numbers representable as the sum of two squares comprise every square, every prime $4k + 1$, a product of such numbers, and the double of any of the preceding.

1636   The great French number theorist Pierre de Fermat announced 17,296 and 18,416 as a pair of amicable numbers. It has been established, however, that this pair of amicable numbers had been previously found by the Arab al-Banna (1256-1321).

1638   French mathematician and philosopher Rene Descartes found the third pair of amicable numbers.

1638   Rene Descartes produced a list of multiply perfect numbers which he could not have discovered without great effort and ingenuity. Although Descartes' fame rests mainly on his philosophical method and in mathematics on his creation of analytical geometry and the invention of coordinate systems, he was also greatly interested in number theory and made various contributions to it.

1640   French mathematician and lawyer Pierre de Fermat became Europe's finest mathematician and he wrote over 3000 mathematical papers and notes. Fermat is considered the founding father of number theory as a systematic science. He discovered many results in the theory of numbers. Many of his theorems, mostly without proof, were stated in the margin of a translated work of Diophantus. Most of these theorems eventually were proved.

1640   Pierre de Fermat found that a prime of the form $4n + 1$ can be represented as the sum of two squares, i.e., $29 = 25 + 4$. The first published proof of this theorem was given by Leonhard Euler in 1754.

Blaise Pascal (1623-1662), mathematician, physicist, religious philosopher and writer, was the co-founder of the modern theory of probability. At the age of 17 Pascal published an essay on mathematics that was highly regarded in the academic community and praised by Rene Descartes. He invented an early calculating machine to assist his mathematician father in local administration. Pascal's calculating machine served as a starting point in the development of the mechanical calculation that has become so important in our time. He also wrote (1653) so extensively on the triangular arrangement of the coefficients of the powers of a binomial, which had already attracted the attention of various writers, that this arrangement has since been known as Pascal's Triangle. In connection with Pierre de Fermat he laid the foundation for the theory of probability. He also perfected the theory of the cycloid and solved the problem of its general quadrature. Having already, at the age of twenty-five, made for himself an impressive reputation in mathematics and physics, he suddenly determined to abandon these fields entirely and to devote his life to the study of philosophy and religion.

1640 In October, Pierre de Fermat gave the theorem known as the *little Fermat theorem* in a letter to Frenicle de Bessy. The first published proof of this theorem was given by Leonhard Euler in 1736.

1641 Mathematician Frenicle de Bessy, an official at the French mint, proposed the idea that two different representations of a number as a sum of two squares may serve to factor it. However, Leonhard Euler, for whom the method is usually named, seems to have been the first to put it to extensive use.

1642 Blaise Pascal invented an early calculating machine (for addition and subtraction).

1643 Pierre de Fermat developed a method of finding the factors of a given number.

1644 Father Marin Mersenne reported that there were twelve values of n less than 258 which yield Mersenne primes ($M_n = 2^n - 1$).

1650 English mathematician John Wallis came up with an expression for expressing π.

1654 Blaise Pascal used what is known as Pascal's triangle to get the binomial coefficients. Pascal also recognized that the binomial coefficients were the same as the formula for the number of combinations of n things taken k at a time.

1655 John Wallis used algebraic notation very similar to what is currently used, including negative exponents and the symbol currently used for infinity.

1658 William Brouncker expressed π in the form of an infinite fraction.

1659 Swiss mathematician Johann H. Rahn introduced the $\div$ sign for division.

1659 Johann H. Rahn published a factor table containing the numbers up to 24,000 excluding those divisible by 2 and 5.

1665 Sir Isaac Newton calculates π to at least 16 decimal places (not published until 1737).

1666 Sir Thomas Morland developed an operational multiplier.

Pierre de Fermat (1601-1665) was a leading mathematician of the 17th century. Fermat, who was a jurist, made mathematics an avocation. He was without a peer for about a century in the modern theory of numbers, and he discovered analytic geometry independently of Rene Descartes. He is also regarded as the discoverer of Fermat's numbers and Fermat's Last Theorem. **Fermat prime numbers** are prime numbers (those not divisible by an integer larger than one except themselves) of the form $2^x + 1$, in which x is 2^n, n being an integer. On the basis of his knowledge that numbers of this form are prime for values of n from 1 through 4, Fermat conjectured that all numbers of this form are prime. An 18th-century Swiss mathematician, Leonhard Euler, showed this conjecture to be false. **Fermat's Last Theorem** is the statement that there are no natural numbers x, y, and z such that $x^n + y^n = z^n$, in which n is a natural number greater than 2. About this Fermat wrote in 1637 in his copy of Claude-Gaspar Bachet's *Diophanti*, "I have discovered a truly remarkable proof but this margin is too small to contain it."

1667   Scottish mathematician James Gregory developed the arctangent series which has played so great a part in calculations of π.

1668   English mathematician John Pell extended Johann H. Rahn's factor table to 100,000. For a considerable period this table was the only one available and it was reprinted several times in other works.

1670   Fermat's marginal notes about his last theorem were published by his son and some of Fermat's letters appeared in Wallis' *Opera Mathematica*.

1671   James Gregory, a Scottish mathematician, discovered an infinite series for the arctangent.

1674   German philosopher and mathematician Gottfried Leibniz discovers the arctangent series for π.

1674   German mathematician and philosopher Gottfried Wilhelm von Leibniz represented π as the limit of an infinite series:
$$\pi = 4(1 - 1/3 + 1/5 - 1/7 + 1/9 - 1/11 + ...).$$

1685   John Wallace realized that logarithms could be thought of as exponents.

1686   Adamas Kochansky, a Pole, succeeded in extending magic squares to three dimensions.

1687   De la Loubere, when envoy of Louis XIV to Siam, determined a simple method for finding a normal magic square of any odd order.

1693   Seki Kowa solved systems of linear equations by representing the coefficients with bamboo sticks arranged in square patterns. Kowa was the most famous Japanese mathematician of the seventeenth century. He is noted for his work on the theory of magic squares and magic circles. Magic squares originated in China many years earlier, but Seki Kowa discovered patterns enabling him to write magic squares by formula rather than trial and error.

1693   Bernard Frenicle de Bessy found that there are 880 magic squares (excluding rotations and reflections) of order 4.

Gottfried Leibniz (1646-1716), a German mathematician and philosopher, thought it was unworthy of excellent men to lose hours like slaves in the labor of calculations which could safely be relegated to anyone else if machines were used. Leibniz made considerable contributions to mathematicians but is unfortunately often remembered most vividly for the controversy over the "Invention" of the calculus. But is is his notation for the calculus, which he introduced in 1675, which has survived. Even though, Leibniz had his calculus theory well developed in 1675, it was not until 1684 that he published, in the *Acta Eruditorum*, a description of the method and its possibilities.

1694 Johann Bernoulli realized that logarithms could be thought of as exponents.

1698 The dot for multiplication was introduced by German mathematician Gottfried Leibniz.

1705 Astronomer Abraham Sharp used an arcsine series to obtain π to 72 decimal digits.

1706 John Machin used the difference between two arctangents to find π to 100 decimal places.

1706 The symbol π, for the circle ratio, used by William Jones.

1710 Thomas Fuller, a black slave in Virginia, could mentally multiply four- and five-digit numbers and calculate the number of seconds in any period of time.

1713 Chinese Emperor Kangshi published a table of logarithms.

1714 Roger Cotes in his work *Logometria* computes the base of natural logs to be 2.7182818.

1719 The French mathematician DeLagny calculates π to 112 correct places.

1719 The first book on mathematics was printed in America: James Hoddler's *Arithmetick*, printed in Boston. It was the first math textbook to appear in English on this side of the Atlantic.

1722 Japanese mathematician Takebe, using a 1024-sided polygon, found π to 41 decimal places.

1727 Swiss scientist Leonhard Euler introduced the symbol e for the base of natural logarithms.

1727 John Hill found that $11,826^2$ is the smallest pandigital square.

Swiss scientist Leonhard Euler (1707-1783) was the most prolific mathematician of the 18th century and perhaps of all time. Euler did for modern analytic geometry and trigonometry what the *Elements* of Euclid had done for geometry, and the resulting tendency to render mathematics and physics in arithmatical terms has continued ever since. Euler made decisive and formative contributions to geometry, calculus, and number theory. Euler enriched mathematics with many concepts, techniques, and notations currently in use. His discoveries in mathematics are so numerous that his collected work will eventually fill about 80 volumes. All his life, Euler worked intensively on problems in number theory.

1729 *Arithmetick, Vulgar and Decimal*, an early arithmetic book printed in Boston. It was written by Isaac Greenwood who held for some years the chair of mathematics in what was then Harvard College.

1732 Leonhard Euler proved that Fermat number F_5 is composite, and since then many more values of F_n have been shown to be composite: $F_5 = 2^{32} + 1 = 4,294,967,297 = 641 \times 6700417$.

1736 Leonhard Euler proved Fermat's minor theorem.

1737 The symbol π was used by the early English mathematicians William Oughtred, Isaac Barrow, and David Gregory to designate the circumference, or periphery, of a circle.

1739 Swiss scientist Leonhard Euler showed that any factor of a Fermat number must have the form $2^{t+1} k + 1$.

1739 Japanese mathematician Matsanuga, using a series, found π equal to 50 decimal places.

1739 The symbol for the circle ratio (π) was first used in 1706 by William Jones who occasionally edited and translated (from Latin) some of Sir Isaac Newton's works, and who himself wrote on general mathematics. However, Jones was not able to make his notation generally accepted. Thirty-three years later, Leonhard Euler used the notation and it became a standard symbol.

1742 German mathematician Christian Goldbach, in letters to Leonhard Euler, conjectured that every even integer is the sum of two primes, and that every odd number is the sum of three primes.

1745 Leonhard Euler mentioned in a letter to Christian Goldbach the factorization by means of representation of a number as the sum of two squares.

1746 Kruger computed a table of primes up to 100,000.

1747 Swiss mathematician Leonhard Euler undertook a systematic search for amicable numbers and produced a list of thirty pairs, which he later extended to more than sixty.

1747 Leonhard Euler showed that any factor of a Fermat number F_n is of the form $k \times 2^{n+1} + 1$.

Joseph Louis Lagrange (1736-1813) was a mathematician who excelled in all fields of analysis and number theory. Lagrange taught mathematics while still in his teens. By 1761 he was recognized as the greatest living mathematician. When he was twenty-three years old he published two memoirs which at once attracted attention. In 1776, Frederick the Great wrote that "the greatest king in Europe" wanted "the greatest mathematician in Europe" in his court. As a result of this letter, Lagrange went to Berlin and remained there more than twenty years. It is probable that his mathematical work more profoundly influenced later mathematical research than did that of any of his contemporaries, although it was an era of giants in this field.

1748 Leonhard Euler publishes the *Introduction in analysin infinitorum*, containing Euler's Theorem and many series for π and π^2.

1748 Maria Agnesi published a widely used math text covering topics from algebra through calculus.

1749 Leonhard Euler defined algebraic functions and used *e* for the base of natural logarithms.

1750 Leonhard Euler proved that Mersenne number M_{31} is prime; $2^{31} - 1$ was the 8th known Mersenne prime.

1751 Calculating prodigy Jedediah Buxton calculated mentally the number of cubic inches in a block of stone 23,145,789 yards long, 5,642,732 yards wide, and 54,965 yards thick.

1752 In November, Christian Goldbach expressed, in a letter to Leonhard Euler, his belief that every odd integer could be written in the form $p + 2a^2$, where p is a prime (or 1, then considered a prime) and $a \geq 0$ is an integer.

1753 Leonhard Euler found that Fermat's Last Theorem is true for $n = 3$.

1754 Jean Etienne Montucla, an early French historian of mathematics, wrote a history of the quadrature problem.

1755 Leonhard Euler derives a very rapidly converging arctangent series for π.

1757 Swiss mathematician Leonhard Euler proved that every every perfect number must be of Euclid's form: $2^{n-1}(2^n - 1)$. It has been proved also that every even perfect number must end in 6 or 28; if it ends in 6, the digit preceding it must be odd.

1767 Johann Lambert proved that π is an irrational number. It is a nonrepeating, nonterminating decimal. Its exact value cannot be computed.

1770 French mathematician Joseph Louis Lagrange published the first proof that every positive integer can be written as the sum of at most four squares. Bachet made this conjecture earlier and verified it for all numbers up to 120.

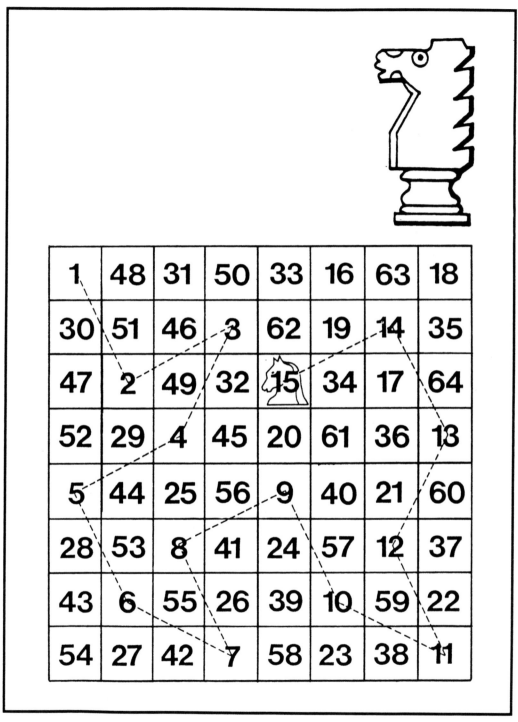

1	48	31	50	33	16	63	18
30	51	46	3	62	19	14	35
47	2	49	32	15	34	17	64
52	29	4	45	20	61	36	13
5	44	25	56	9	40	21	60
28	53	8	41	24	57	12	37
43	6	55	26	39	10	59	22
54	27	42	7	58	23	38	11

An 8 x 8 magic square made by Swiss mathematician Leonard Euler. It can be split into four separate magic squares. You can also move a chess knight across the grid without visiting a square more than once.

1770   Johann Heinrich Lambert developed a table of least factors of numbers up to 102,000. He also proved the irrationality of π.

1770   Leonhard Euler, a remarkable blind scientist published *Complete Introduction to Algebra*, a textbook which contained much material on elementary number theory.

1770   English mathematician Edward Waring published his *Meditationes algebraicae* which contained several announcements and conjectures on the theory of numbers, among them the fact that every number can be represented as the sum of a limited number of cubes, fourth, or higher powers. He also stated without proof that every positive integer can be expressed as the sum of nine cubes and as the sum of 19 fourth powers.

1771   Leonhard Euler lost the sight of his remaining eye, but dictating first to his children and later to a secretary, the flood of his mathematical research and publications continued unabated. Euler, totally blind, wrote many works such as his famous treatise on celestial mechanics.

1772   Leonhard Euler gave the formula $n^2 - n + 41$, which yields a prime number for all integral values of n from 0 to 40.

1775   Leonhard Euler suggests that π is transcendental.

1776   German scientist Johann Heinrich Lambert published a factor table containing numbers up to 408,000. The tables were published at the expense of the Austrian imperial treasury, but since there was a disappointing number of subscribers, the treasury confiscated the whole edition except a couple of copies, and the paper was used in cartridges in a war against the Turks.

1776   Antonio Felkel calculated the prime factors of all numbers to 2,000,000.

1777   Comte de Buffon devised his famous "needle problem," by which π may be approximated by probability methods.

1794   Adrien-Marie Legendre, in his *Elements de Geometric* proved the irrationality of π more rigorously, and also gave a proof that π^2 is irrational.

Swiss mathematician Carl Friedrich Gauss (1777-1855), along with Archimedes and Newton, is considered one of the three greatest mathematicians of all time. While still in grade school, he discovered and proved the formula $1+2+3+...+n=n(n+1)/2$. By the time he was 21 he had contributed more to mathematics than most mathematicians do in a lifetime. He is especially famous for his pioneering work in the theory of numbers. Gauss called *number theory* "the queen of mathematics." He also called mathematics "the queen of the sciences."

1794   Vega using a new series for the arctangent discovered by Leonhard Euler, calculated π to 140 decimal places.

1796   Carl Friedrich Gauss proved that every integer is the sum of at most three triangular numbers. He made this entry in his diary, dated July 10, when he was only 19 years of age.

1797   Vega calculated a table of primes up to 400,031.

1799   The Napoleonic expedition to Egypt found a trilingual tablet at Rosetta near Alexandria, the Rosetta stone. Its message was recorded in Greek, Demotic and Hieroglyphic.

1799   David Rittenhouse made America's earliest contribution to mathematics research when he wrote on methods of computing with logarithms.

1801   German mathematician Carl Friedrich Gauss among other things took up the ancient problem of finding all regular polygons that can be constructed by means of compass and ruler.

1801   Carl Friedrich Gauss published his most important work on the properties of numbers, *Disquisitiones arithmeticae*, when he was twenty-four years of age. Six years later, a French translation, *Recherches arithmetiques* was published in Paris by A.C.M. Poullet-Delisle.

1811   Barlows *Number Theory* gives the perfect numbers up to P_{31} corresponding to the Mersenne prime M_{31} obtained by Leonhard Euler, at the time the greatest prime known. This perfect number "is the greatest that will ever be discovered, for, as they are merely curious without being useful it is not likely that any person will attempt to find one beyond it." The great efforts expended since that time in such computations show that it is difficult to underestimate human curiosity. Many future efforts were made possible by the invention of the computer.

1811   Chernac developed a table of prime factors of numbers up to 1,020,000.

1813   Richard Wagner's name contains 13 letters. He was born in 1813 and $1 + 8 + 1 + 3 = 13$. He composed 13 great works of music. *Tannhauser*, one of his greatest works, was completed on April 13, 1845, and it

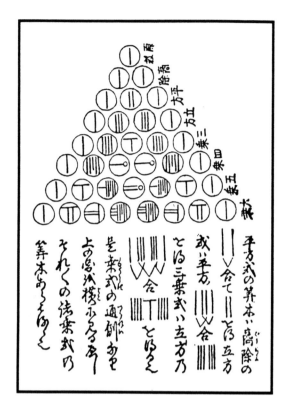

The Japanese version of Pascal's Triangle from Murai Chuzen's Sampo Doshi-mon (1781). This illustration also shows the Sangi board (counting board ruled in columns) forms of the numbers.

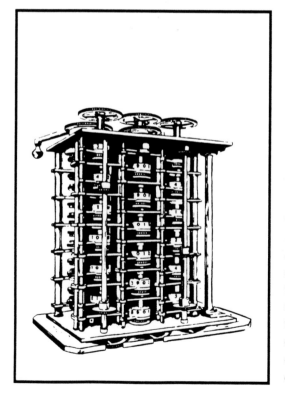

The idea of mechanically calculating mathematical tables first came to Charles Babbage in 1812. In 1822 he designed the Difference Engine, a calculator that could tabulate certain mathematical computations to eight decimals. Babbage later designed another calculator, called the Analytical Engine, with a 20-decimal capacity that was capable of computing a variety of more complex problems.

was first performed on March 13, 1861. He finished *Parsifal* on January 13, 1842. *Die Walkure* was first performed in 1870 on June 26, and 26 is twice 13. *Lohengrin* was composed in 1848, but Wagner did not hear it played until 1861, exactly 13 years later. He died on February 13, 1883; the first and last digits of this year form 13.

1816   The French Academy announces a prize for a solution to Fermat's Last Theorem.

1816   Burkhardt computed a table of least factors of numbers up to 3,036,000.

1817   The *Canon Pellianus* computed by C.F. Degen gave the solution of $y^2 - Cx^2 = 1$ for all non-squares, C, not exceeding 1000.

1819   It was discovered that $2^{340} \equiv 1 \pmod{341}$. As $341 = (11)(31)$ is thus a composite number satisfying the Fermat condition (Fermat's Theorem states that certain powers upon division by primes always yielded remainders of 1), a new class of numbers came to light. They are called *pseudo-primes*.

1821   Russian mathematician P.L. Tchebychev proved that if $n > 1$ than there is always at least one prime between n and $2n$.

1822   Englishman Charles Babbage designed a *Difference Engine* for the calculating and checking of mathematical tables.

1823   Henry Goodwyn published in London a table 107 pages long, giving the decimal periods of every fraction having a prime or composite denominator that is relatively prime to 10 and not exceeding 1024.

1824   Johann Dase of Germany was a professional calculator by the age of 15. He could mentally multiply two 20-digit numbers in 6 minutes and extract the square root of a number with 100 digits in 52 minutes.

1824   Wilham Rutherford, an Englishman, calculated π to 208 decimal places, of which 152 were later found to be correct.

1825   French mathematician Sophie Germain made a major contribution toward proving Fermat's Last Theorem. In her honor, a prime P, whose companion $Q = 2P + 1$ is also prime, is called a Sophie Germain Prime.

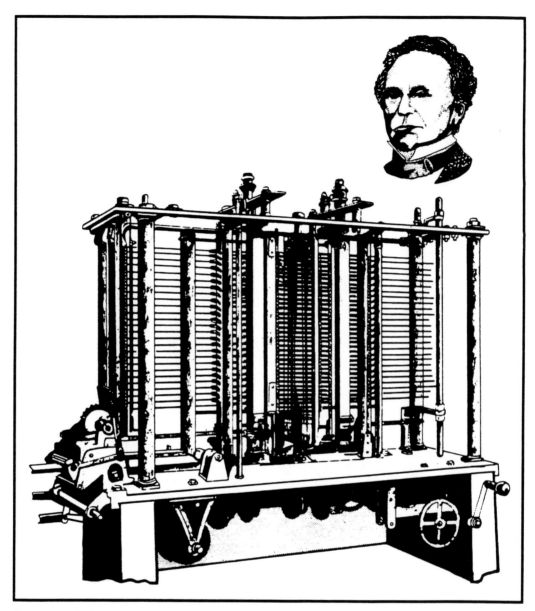

Charles Babbage (1792-1871), an English mathematician and mechanical genius, devoted 37 years to perfecting a calculating machine that could be used to construct mathematical tables. Although his project did not succeed because of inadequate technology, his underlying logic was sound, leading to the modern day computer. Babbage's ideas were so advanced and his standards so high, that most efforts to realize his plans during his own lifetime were unsuccessful. Babbage's machine was called the Analytical Engine which he conceived in 1833. The machine would calculate, it would process statistics, and it would automatically guide its own actions based on the answers it was producing.

1825 P.G.Lejeune Dirichlet and A.M. Legendre found that Fermat's Last Theorem is true for $n = 5$.

1832 Carl Friedrich Gauss initiated the use of term "complex number" for the sum of a real number and an imaginary number.

1836 Truman Safford of Vermont mentally raised 365 to the 11th power in no more than a minute.

1837 William Rowan Hamilton showed that complex numbers can be considered as ordered couples of real numbers.

1837 P.G. Lejeune Dirichlet showed that every set of integers of the form $an + b$, where a and b are relatively prime integers, contains an infinite number of primes. Thus the set of integers of the form $3n + 1$, that is, the set 1, 4, 7, 10, 13, ..., contains infinitely many primes, as does the set of the form $6n + 5$; 5, 11, 17, 23, ..., and so forth.

1839 K.G.J. Jacobi published an extensive table of primitive roots for primes for all primes below 1000.

1840 Gabriel Lame proved that Fermat's equation, $x^7 + y^7 = z^7$ has no solutions in integers.

1840 J.P.M. Binet derived a formula for finding the value of any Fibonacci number, given that its place in the sequence is known.

1840 Joseph Liouville proves the existence of transcendental numbers. He showed that one could define numbers which cannot be roots of any algebraic equation.

1841 Karl Weierstrass introduced the absolute value symbol.

1843 German mathematician E. Kummer extended the domain of number theory to include not only the rational numbers but also the algebraic numbers, i.e., numbers satisfying algebraic equations with rational coefficients.

1844 Liouville proved the existence of transcendental numbers. Four years later, Canto later proved that almost all numbers are transcendental.

George Boole (1815-1864), an English mathematician and logician, was one of the first to direct attention to the theory of invariants. This concerns expressions in several variables that do not change when the coordinates change. Boole also did important work on finite differences and differential equations. This led him to think of differential operators algebraically, and gradually he was led to consider the operations of logic algebraically also. This led him to the work for which he is best remembered, his *Mathematical Analysis of Logic* (1847) and *Investigation of the Laws of Thought* (1854). In these he employed mathematical symbolism to express logical relations, thus becoming an outstanding pioneer of modern symbolic logic. *Boolean algebra* is a generalization of the familiar operations of arithmetic to quantities that obey rules analogous to those Boole proposed for sets; it is particularly useful in the design of circuits and computers.

1844   Viennese mathematician L.K. Schulz von Strassnitsky and calculating prodigy, Johann Martin Zacharias Dase calculated π correct to 200 places in less than two months.

1845   J. Bertrand stated that for any integer n greater than 3, there is a prime between n and $2n$.

1847   Calculating prodigy Zacharias Dase, a German human-calculator, reckoned out the natural logarithms (7 places) of the numbers from 1 to 1,005,000. He performed many other calculating feats including extracting mentally the square root of a number of 100 figures in 52 minutes.

1847   Thomas Clausen found π to 248 decimal places.

1847   E.E. Kummer found that Fermat's Last Theorem is true for $n < 37$.

1851   French mathematician Joseph Liouville found the first transcendental number. The decimal form of Liouville's number is
        .110001000000000000000000100...

        and has a one in the 1st, 2nd, 6th, 24th, 120th, etc. places and zeros elsewhere. Liouville succeeded in finding a whole class of numbers that were all transcendental.

1852   Francis Guthrie conjectured that four colors are sufficient to color all maps so that countries with at least an arc as a common boundary have different colors.

1853   Rutherford returned to the problem and computed π to 440 correct decimal places.

1854   English mathematician George Boole expanded and clarified his earlier work of 1847 into a book entitled *Investigation of the Laws of Thought*, in which he established both formal logic and a new algebra — the algebra of sets known today as Boolean algebra.

1855   Richter calculates π to 500 decimal places.

1856   Crelle presented to the Berlin Academy a table of primes up to 6 million. This work was extended by Zacharias Dase, before 1861, up to 9 million, entirely done by mental calculation.

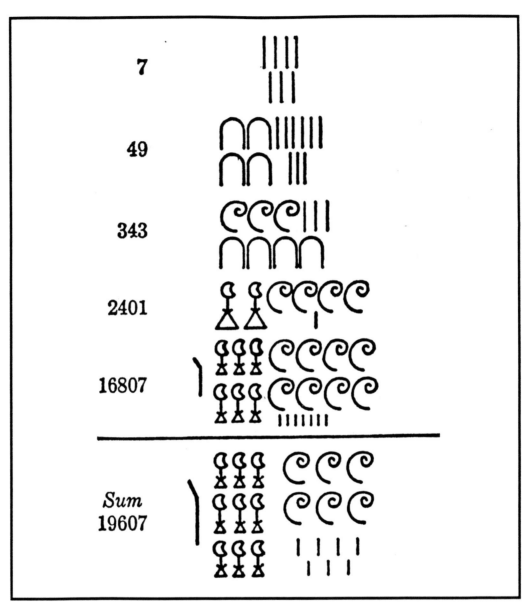

In 1650 B.C. a scribe named Ahmes copied an older treatise on mathematics that contained 84 problems and their solutions. It was called the Ahmes Papyrus. The actual manuscript of Ahmes has come down to us, having been purchased in Egypt about the middle of the 19th century by the English Egyptologist A. Henry Rhind, and having later been acquired by the British Museum. It is one of the oldest mathematical manuscripts on papyrus extant. The Ahmes Papyrus is a practical handbook. It contains material on linear equations; it treats extensively of unit fractions; and it includes problems. The Ahmes Papyrus is often called the Rhind Papyrus. Shown here is a problem from the papyrus.

1857 E.E. Kummer develops complicated criteria for proving Fermat's Last Theorem for certain irregular primes. The French Academy presented Kummer with a gold medal valued at 3,000 francs in recognition of his researches in complex numbers.

1857 Arthur Cayley formulated the algebra of matrices.

1858 The oldest mathematical document, Ahmes Papyrus, was found at Thebes in a room of a ruined building and bought by a Scottish antiquary, A. Henry Rhind, in a Nile resort town. Four years later, it was purchased from his estate by the British Museum. The Ahmes Papyrus contains 84 problems and their solutions. It is often called the Rhind Papyrus.

1860 Baldassare Boncompagni published the 1228 version of Leonardo Fibonacci's *Liber Abaci*.

1863 Kulik produced a factor table of numbers to 100,330,200 (except for multiples of 2, 3, 5). He spent about 20 years preparing this table.

1865 English mathematician Lewis Dodgson (Lewis Carroll) used his story *Alice in Wonderland* to amuse children and poke fun at human frailties. He wrote many books and papers concerning mathematics.

1866 A sixteen-year old Italian boy Nicolo Paganini found the amicable number pair 1184 and 1210. This relatively small pair of amicable numbers had been overlooked by Leonhard Euler and other mathematicians.

1866 By now, 65 pairs of amicable numbers were known, three discovered before Leonhard Euler, 59 discovered by Euler, and three discovered after Euler.

1868 The Sator stone was found in Ciencester, England, but it is believed to be much older, because an identical stone was excavated from the ruins of Pompeii. The Sator stone is something like an alphabetical magic square.

1870 Ernst Meisel showed that the number of primes below 10^8 is 5,761,455.

1873 French mathematician Charles Hermite showed that e, the base of the natural logarithms, is a transcendental number.

```
100 REM ****************************
110 REM **** PRIME POLYNOMIAL ****
120 REM ****************************
130 REM **** P — PRIME NUMBER ****
140 REM **** X — LOOP AND VALUE OF X ****
150 PRINT "PRIME NUMBERS"
160 FOR X = 1 TO 40
170     LET P = X ∧ 2 - X + 41
180     PRINT P,
190 NEXT X
200 END

RUN

PRIME NUMBERS
41        43        47        53        61
71        83        97        113       131
151       173       197       223       251
281       313       347       383       421
461       503       547       593       641
691       743       797       853       911
971       1033      1097      1163      1231
1301      1373      1447      1523      1601
```

For centuries, mathematicians have tried to find a formula that would yield every prime — or even a formula that would yield only primes. No one has yet found such a formula, but several remarkable expressions produce large numbers of primes for consecutive values of x. For example, $2x^2 + 29$ will give primes (starting with 29) for x = 0 to 28 (twenty-nine primes), $x^2 + x + 41$ will give primes for x = 0 to 39 (forty primes starting with 41); $x^2 + x + 17$ will generate sixteen primes, $6x^2 + 6x + 31$ will give primes for twenty-nine values of x, $3x^2 + 3x + 23$ will give primes for twenty-two values of x, and $x^2 - 79x + 1601$ will give eighty consecutive prime values when x = 0,1, 2, ..., 79. Other examples of the same nature exist.

The above BASIC program uses the formula

$$x^2 - x + 41$$

to generate primes for the 40 values of x: 1, 2, 3, ..., 40.

1873 A British mathematician named William Shanks worked out π to 707 decimal places. However his calculations contained an error after 527 correct decimals.

1874 Repunit numbers have a simple relationship to powers of 10: $R_n = (10^n - 1)/9$. William Shanks, the calculator of π, developed the first table of prime repunit numbers as powers of 10 as an aid to finding a prime from the length of the period of its decimal reciprocal.

1876 French mathematician Edouard Lucas determined that
$$170,141,183,460,469,231,731,687,303,715,884,105,727$$
is a large Mersenne prime. This 39-digit number was the largest prime known before the age of computers. Lucas discovered a fast way to test for the primality of a Mersenne number. He later expressed some doubt about this result but it was confirmed in 1914 by E. Fauquembergue.

1876 Edouard Lucas discovered that the sums of the numbers on the rising diagonals of the Pascal Triangle form the Fibonacci sequence.

1876 Edouard Lucas invented a Fibonacci-like number sequence that bears his name.

1879 E.B. Escott gave the formula $n^2 - 79n + 1601$, which gives primes for all $n = 0, 1, 2, \ldots, 79$. It has been proved that no such polynomial function can ever produce only primes.

1880 Gottfried Ruckle of Germany could recite all the factors of each integer less than 1000 by the age of 12.

1880 F. Landry showed that Fermat number $F_6 = 2^{64} + 1$ is the product of two primes: 274,177 and 67,280,421,310,721. However, Pervusin had already discovered that Fermat number F_{12} is divisible by $7 \times 2^{14} + 1 = 114,689$. The Fermat numbers, like the Mersenne numbers, had become an ideal testing ground for primality tests and methods of factorization.

1882 German mathematician F. Lindemann showed that π is a transcendental number. Lindemann's proof was 13 pages of tough mathematics.

1883 I.M. Pervushin discovered the 9th Mersenne prime, $2^{61} - 1$.

```
3.141592653589793238462643383279502884197169
3993751058209749445923078164062862089986280 3
4825342117067982148086513282306647093844609 5
5058223172535940812848111745028410270193852 1
1055596446229489549303819644288109756659334 4
6128475648233786783165271201909145648566923 4
6034861045432664821339360726024914127372458 7
0066063155881748815209209628292540917153643 6
7892590360011330530548820466521384146951941 5
1160943305727036575959195309218611738193261 1
7931051185480744623799627495673518857527248 9
1227938183011949129833673362440656643086021 3
9494639522473719070217986094370277053921717 6
2931767523846748184676694051320005681271452 6
3560827785771342757789609173637178721468440 9
0122495343014654958537105079227968925892354 2
0199561121290219608640344181598136297747713 0
9960518707211349999998372978049951059731732 8
1609631859502445945534690830264252230825334 4
6850352619311881710100031378387528865875332 0
8381420617177669147303598253490428755468731 1
5956286388235378759375195778185778053217122 6
8066130019278766111959092164201989380952572 0
1065485863278865936153381827968230301952035 3
0185296899577362259941389124972177528347913 1
5155748572424541506959508295331168617278558 8
9075098381754637464939319255060400927701671 1
3900984882401285836160356370766010471018194 2
9555961989467678374494482553797747268471040 4
7534646208046684259069491293313677028989152 1
0475216205696602405803815019351125338243003 5
```

1883 After Lindemann's proof of 1882, the activity of computing π ceased. The computation of π went to sleep once again.

1885 Karl Wilhelm Weierstrass simplified the 1883 proof of Lindemann's theory that π is a transcendental number.

1888 Beginnings of the American Mathematical Society.

1892 Richelot showed how to construct a regular 257-gon. Carl Gauss had proved that a regular polygon with a prime number of sides can be constructed only if that number is a Fermat prime.

1893 Danish mathematician Bertelsen announced that the number of primes below 10^8 is 50,847,478. In 1959 the American mathematician D.H. Lehmer showed that this result is incorrect and that it should read 50,847,534.

1895 Alexander Aitken of New Zealand memorized the first 1000 digits of π.

1896 Salo Finkelstein of Russia could memorize 30 digits of a number after looking at it for only 3 seconds.

1896 German mathematician G.F.B. Riemann is regarded as the founder of analytic number theory. On the basis of his ideas, the prime number theorem was proved independently by the French mathematician J. Hadamard and the Belgian C.J. de la Vallee-Poussin.

1896 Pal Erdos' most famous proof is of the prime number theorem, which says that if $\pi(x)$ is the number of primes not exceeding x, then x tends to infinity, $(\pi(x) \log x)x$ tends to 1. It was proved using complex analysis (using complex numbers). In 1949, Erdos published a proof that avoided complex numbers entirely. Such a proof is called "elementary," however, "elementary" does not mean easy, merely that complex numbers are not used.

1897 The Indiana House of Representatives passed a bill that attempted to legislate the value of π. The author of the bill was a physician, Edwin J. Goodman, M.D. from Posey County. It was introduced in the Indiana House on January 18 by Posey County Representative Taylor I. Record. It was entitled "A bill introducing a new Mathematical truth," and it became House Bill No. 246. On February 5, the House

$$M_{67} = 2^{67} - 1 =$$
$$(193707721) \times$$
$$(761838257287)$$

At the October 1903 meeting of the American Mathematical Society, the mathematician Frederick Nelson Cole was listed in the program as presenting a paper with the title "On the factorization of large numbers." When called upon to speak, Cole walked up to the blackboard and, without uttering a word, performed the calculation of 2 raised to the power of 67, following which he subtracted 1 from the result. Still saying nothing he moved to a clean part of the board and multiplied together the two numbers 193,707,721 and 761,838,257,287. The answers to the two calculations was identical. Cole resumed his seat still having said not one word, and for the first and only time on record, the entire audience at an American Mathematical Society meeting rose and gave a "speaker" a standing ovation. What Cole had done was to find the prime factors of the Mersenne number M_{67}.

passed the learned treatise unanimously. However, on February 12, a wiser, more informed Senate, voted to postpone further consideration of the bill.

1903   F.N. Cole proved M_{67} to be a composite number. (Father Marin Mersenne had believed M_{67} to be prime). He was later asked how long it had taken him to find this factorization. Cole replied, "Three years of Sundays."

1903   Western found one factor of Fermat number F_9.

1904   The great German mathematician David Hilbert presented a list of 24 problems which he thought were worthy of the attention of twentieth century mathematicians.

1904   R. Chartres reported an application of the known fact that if two positive integers are written down at random, the probability that they will be relatively prime is $6/\pi^2$.

1905   The first magic cube was privately published.

1906   American mathematician A.C. Orr created a method where the numbers of letters in each word gives the digits of π, e.g., May I have a large container of orange juice.

1908   German mathematician F.P. Wolfskehl bequeathed 100,000 marks to the Academy of Science in Gottingen for a prize to be awarded for the first complete proof of Fermat's Last Theorem. Over 1000 false "complete" proofs were presented during the period from 1908 to 1911.

1909   Wieferich proved that Fermat's equation $x^p + y^p = z^p$, has a solution in which p is an odd prime that does not divide any of x, y or z, then $2^{p-1} - 1$ is divisible by p^2.

1909   D.H. Lehmer published a factor table for the first ten millions containing the smallest factor of every number not divisible by 2, 3, 5 and 7 between the limits 0 and 10,017,000.

1909   Moorhead and Western proved that Fermat numbers F_7 and F_8 are composite, without producing any factors. In 1970, Morrison and Brillhart found the two prime factors for $F_7 - (2^9 \times 116,503,103,764,643 + 1)$ and$(2^9 \times 11,141,971,095,088,142,685 + 1)$. F_8 was finally

The brilliant Indian mathematician Srinivasa Ramanujan is pictured on this postage stamp. He developed many results in number theory. He specialized in the study of numbers and knew their characteristics in the same way that a baseball fan might know a vast number of statistics about the game. The late G.H. Hardy (1877-1947), a leading mathematician of his time, once rode in a taxicab to see his ill Indian protege, Ramanujan. The number of the taxicab was 1729, and Hardy said "that the number seemed to me rather a dull one, and that I hoped it was not an unfavourable omen. 'No,' Ramanujan replied, 'it is a very interesting number; it is the smallest number expressible as a sum of two cubes in two different ways.'"

$$1729 = 12^3 + 1^3 = 1728 + 1$$
$$1729 = 9^3 + 10^3 + 729 + 1000$$

conquered in 1981 when R.P. Brent and John Pollard found the prime factor 1,238,926,361,552,897.

1909 German mathematician David Hilbert proved a 1770 conjecture by Edward Waring, that every number can be represented as the sum of a limited number of cubes, fourth, or higher powers.

1909 9^{9^9} is the largest number in decimal notation that can be represented without using more than 3 digits, with no additional symbols. C.A. Laisant showed that this number has 369,693,100 digits.

1911 R.E.Powers discovered the 10th Mersenne prime, $2^{89} - 1$.

1912 William Klein of Holland mentally multiplied two five-digit numbers in 44 seconds.

1912 A. Martin published a list of all primitive integral Pythagorean triangles for which the hypotenuse does not exceed 3000: *Proceedings*, Fifth International Mathematical Congress, Cambridge.

1912 R.D. Carmichael proved that every Carmichael number is the product of at least 3 odd primes.

1913 Srinivasa Ramanujan, who possessed amazing ability to see quickly and deeply into intricate number relations, was discovered by the eminent British number theorist, G.H. Hardy. A most remarkable mathematical association resulted between the two men.

1913 J.N. Muncey of Jessup, Iowa produced a magic square composed of consecutive prime numbers (leaving out the prime 2).

1913 E. Fauquembergue discovered the 11th Mersenne prime, $2^{107} - 1$.

1913 L.E. Dickson proved that the five smallest amicable pairs are (220, 284), (1184, 1210), (2620, 2924), (5020, 5564), and (6232, 6368).

1914 D.H. Lehmer published *List of Prime Numbers from 1 to 10,006,721*, Carnegie Institution of Washington Publication 165.

1914 The ancient Greeks knew the first four perfect numbers and by this year eight more had been discovered.

1915 K.L. Jensen proved that there exists infinitely many irregular primes.

LUCAS-LEHMER TEST

Example that applies the
Lucas-Lehmer Test to $M_5 = 31$.

$$u(1) = 4$$
$$u(2) = (4^2 - 2) \bmod 31 = 14$$
$$u(3) = (14^2 - 2) \bmod 31 = 8$$
$$u(4) = (8^2 - 2) \bmod 31 = 0$$

Hence, M_5 is indeed a prime number.

In 1876, Edouard Lucas discovered a fast way to test the primality of a Mersenne Number. Using the test and calculators, several primes were added to the list of Mersenne Primes. In 1930, D.H. Lehmer published an improved version of Lucas' algorithm. The Lucas-Lehmer test for primality is:

Let $u(1) = 4$.
For i = 1 to p − 2 compute $u(i + 1) = (u(i)^2 - 2) \bmod M_p$.
If and only if $u(p - 1) = 0$, then M_p is a prime.

The "mod M_p" means to keep only the remainder after division by M_p.

1918   Frenchman P. Poulet found that 12496 is the first of a chain of five sociable numbers. The sum of the divisors, excluding itself, of each number is the next number in the chain, the last number preceding the first: 12496; 14288; 15472; 14536; 14264.

1918   The second smallest prime repunit, R_{19}, was discovered by one of the readers of H.E. Dudeney's newspaper puzzle column.

1920   A.H. Church, an Oxford University botanist, discovered that the spirals on sunflower heads corresponded to the numbers in Fibonacci's rabbit sequence.

1926   D.H. Lehmer computed the value of e to 709 decimal places.

1927   Lehmer showed that Mersenne number M_{257} is composite.

1927   The Ahmes Papyrus (usually called the Rhind Papyrus), was translated by P.B. Chase and his colleagues.

1929 e^{π} was proved transcendental.

1929   P. Poulet published an extensive list of amicable numbers: *La Chasse aux nombres*, 2 vols. Brussels.

1929   Hans Eberstark of Austria recited the first 11,944 digits of π. Eberstark's feat was bettered by Hideaki Tomoyori of Japan and, later, by Creighton Carvello, who memorized the first 20,013 digits of π.

1930   D.H. Lehmer published an improved version of Edouard Lucas' algorithm for testing the primality of a Mersenne Number.

1930   H.S. Vandiver found that Fermat's Last Theorem is true for $n < 617$. He used a calculator and improvements to E.E. Kummer's criterion.

1931   Russian mathematician L.G. Schnirelmann showed that every positive integer can be represented as the sum of not more than 300,000 primes.

1931   Shyam Marathe of India could mentally raise single digit numbers up to the 20th power.

TWIN	PRIMES
3	5
5	7
11	13
17	19
29	31
41	43
59	61
71	73
101	103
107	109
137	139
149	151
179	181
191	193
197	199
227	229
239	241
269	271
281	283
311	313
347	349
419	421
431	433
461	463
521	523
569	571
599	601
617	619
641	643
659	661
809	811
821	823
827	829
857	859
881	883

Isn't it exciting to be twin primes?

Two consecutive odd integers that are prime are called *twin primes*. Shown here are several twin primes. In 1942, E.S. Selmer and G. Nesheim published a list of twin primes up to 200,000. In 1989, the large twin primes, $1{,}706{,}595 \times 2^{11235} \pm 1$ with 3389 digits each, were found. In 1994, Harvey Dubner found the largest known twin primes $1{,}691{,}232 \times 1001 \times 10^{4020} \pm 1$.

1934 Alexander Gelfond proved that e^π is transcendental: 23.14069263...

1936 D.H. Lehmer contributed a valuable adjunct to the method of testing primality by means of the converse of Fermat's theorem.

1937 Russian mathematician I. Vinogradoff showed by analytic means that every odd number that is sufficiently large is the sum of three odd primes. Unfortunately, "sufficiently large" means greater than $10^{350,000}$, much too large even for modern computers.

1938 Paul Poulet constructed tables containing the composite numbers n up to 100,000,000 for which the congruence $2^{n-1} \equiv 1 \pmod{n}$ is fulfilled, and for each such n a prime factor is given.

1938 Albert L. Candy estimated that there were 13,288,952 magic squares of order 5. The exact number was not known until 1973 when mathematician Richard Schroeppel, using a Digital Equipment Corp. PDP-11 minicomputer, found that there were 275,305,224 magic squares of order 5.

1939 B.H. Brown found the smallest pair of odd amicable numbers: 12,285 and 14,595.

1940 Shahuntala Devi of India extracted the 23rd root of a 501-digit number in 50 seconds.

1942 A list of twin primes up to 200,000 was published by E.S. Selmer and G. Nesheim.

1944 H.J. Kanold showed that there are no odd perfect numbers below 1.4 $\times 10^{14}$.

1944 The concept of superabundant numbers was created.

1944 To while away time during the occupation in World War II, the French mathematician Paul Poulet worked out the prime decomposition of the Mersenne Number M_{135}, finding $M_{135} = (7)(31)(73)(151)(271)(631)$ $(23,311)(262,657)(348,031)(49,971,617,830,801)$.

1945 The Englishman, D.F. Ferguson finds Shank's calculation erroneous from the 528th place onward.

ENIAC (Electronic Numerical Integrator And Calculator), an early electronic comptuer, was built by J. Presper Eckert and John Mauchly at the Moore School of Electrical Engineering, University of Pennsylvania in 1946. It was used by the U.S. Army to compute trajectories for new weapons. In 1949, ENIAC was used to calculate π to 2037 places in 70 hours. The computer evaluation was programmed in accordance with John Machin's formula $\pi = 16 \arctan(1/5) - 4 \arctan(1/239)$.

1946 E.B. Escott produced a list of the 390 known amicable pairs with the names of their discoverers: "Amicable Numbers," *Scripta Mathematica*, Vol. 12.

1946 D.F. Ferguson publishes π to 620 places, using a desk calculator.

1947 In January, D.F. Ferguson of England computed values of π to 710 places and, in September, 808 places using a desk calculator.

1947 A list of twin primes up to 300,000 was prepared by H. Tietze.

1947 W.H. Mills showed that there exists some real number A such that A^{3^n} gives only primes.

1949 Erdos published a proof of the prime number theorem that avoided complex numbers.

1949 John Wrench, Jr. and Levi Smith, computed π to 1120 decimals. This turned out to be the last effort to compute π on a pre-electronic calculator.

1949 The first computer calculation of π was made in September on ENIAC (Electronic Numerical Integrator and Computer) at the Ballistic Research Laboratories; it calculated π to 2037 places in 70 hours. The computer evaluation was programmed in accordance with Machin's formula $\pi = 16 \arctan (1/5) - 4 \arctan (1/239)$.

1949 D.R. Kaprekar announced that when the digits, not all alike, of any four-digit integer are rearranged to form the greatest and least numerical values, and the same routine is applied to the difference of these numbers, the repetitive routine eventually will produce, in not over seven operations, 6174, now known as Kaprekar's constant.

1949 D.H. Lehmer published a table of composite numbers n, satisfying the congruence
$$2^{n-1} \equiv 1 \ (\bmod \ n)$$
to $n = 200,000,000$ by means of the Army Ordnance ENIAC computer.

1949 Captain Benson constructed a tri-magic square of order 32. This is a square that retains its magic not only when every number is squared but also when every number is cubed. The simplest known square of this type was of order 64, before Captain Benson found a way to half the order.

In August 1950, a versatile and highly trained team developed the National Bureau of Standards' Western Automatic Computer (SWAC). It was the first parallel stored-program computer to become operational. It also was the fastest computer in existence at that time. It was a major development in the history of computing. It was designed at a time when "computers" primarily were calculators with "programs" determined by switch settings. It was produced in a period before there was a computer industry, when computers, or calculators, were built one at a time, usually at the place where they were to be used, and when only one of a kind ever was produced. In 1952, R.M. Robinson, with the assistance of D.H. and E. Lehmer used the SWAC computer to discover five new Mersenne prime numbers, the first such discoveries with a computer.

1950 John Wallis, an English mathematician developed an infinite rational product for π.

1950 D.H. Lehmer discovered the smallest even pseudoprime to base 2: $161,038 = 2 \times 73 \times 1103$. Pseudoprimes are relatively rare, though there are an infinite number of them. The next is 215,326.

1951 In July, J.C.P. Miller and D.J. Wheeler of Cambridge University used the EDSAC computer to find a new 41-digit prime number, $180 \times (2^{127} - 1) + 1$. In the same month, A. Ferrier, using a desk calculator, showed the primality of $(2^{143} + 1)/17$, which was then the second largest known prime.

1951 English computer scientist and mathematician Alan Turing made the first attempt to find Mersenne primes using an electronic computer, however, he was unsuccessful.

1951 Only twelve perfect numbers are known.

1951 Based on calculations by Kulik, Poletti, and Porter, a new prime table was published by N.G. Beeger in Amsterdam.

1952 R.M. Robinson carried out Lucas' test using a SWAC computer (from the National Bureau of Standards in Los Angeles), with the assistance of D.H. and E. Lehmer. He discovered Mersenne primes M_{521} and M_{607} on January 30 — the first such discoveries with a computer. The Mersenne primes M_{1279}, M_{2203} and M_{2281} were found later in the year. For the last two of these Mersenne primes, after coding, the actual running time of the SWAC computer amounted to 59 and 66 minutes, respectively.

1952 With the aid of the SWAC computer, five more perfect numbers were discovered, corresponding to $n = 521, 607, 1279, 2203,$ and 2281 in Euclid's formula.

1952 D.H. Wheeler computed the value of e to 60,000 decimals on the ILLIAC, a computer located at the University of Illinois. It took the machine 40 hours to complete the job.

1953 Goldberg discovered that 563 is a Wilson prime. In 1988, a research team conducted a search for the next Wilson prime up to 10^7 without finding a new prime.

In the fall of 1954 the Watson Scientific Laboratory of the IBM Corporation announced the completion of the Naval Ordance Research Calculator — the NORC. The NORC was five times faster than the IBM 701. The NORC, being a machine with a very high speed, a rather large memory, and a very exceptional capability to ingest data, was suited for problems where large amounts of material have to be processed. In 1955, the NORC was used to calculate π to 3089 decimal digits. It accomplished this task in 13 minutes.

1953 Catalin conjectured that if p is a Mersenne prime, then M_p will be prime. This was proved by Wheeler on the ILLIAC computer in one hundred hours.

1953 A new list of multiply perfect numbers was published by B. Franqui and M. Garcia: *American Mathematical Monthly*, Vol. 60.

1954 A.L. Brown (*Scripta Mathematica*, Vol. 20) added more than 200 new multiply perfect numbers to those previously known.

1954 D.H. Lehmer, E. Lehmer and H.S. Vandiver found that Fermat's Last Theorem is true for $n \leq 2500$.

1955 NORC (Naval Ordnance Research Calculator) at Dahlgren, Virginia, was programmed to calculate π to 3089 decimal places; the run took 13 minutes.

1956 Stanislav Ulam and a group of scientists at Los Alamos Scientific Laboratories discovered lucky numbers.

1957 In March, a Ferranti PEGASUS computer at the Ferranti Computer Centre, London, computed π to 10,021 decimal digits in 33 hours. However, a subsequent check revealed that a machine error had occurred, so that only 7,480 places were correct. The run was therefore repeated in March 1958, but the correction was not published.

1957 A perfect number ($n = 3,217$ in Euclid's formula) was found using the Swedish computer BESK.

1957 H. Riesel discovers the 18th Mersenne prime, $2^{3217} - 1$.

1958 Felton, in England, using a Ferranti PEGASUS computer, calculated π to 10,000 decimal places in 33 hours.

1958 R.M.Robinson showed that Cullen number C141 is a prime, the only known Cullen prime for 25 years, apart from C1 = 3.

1958 D.H. Lehmer and E. Lehmer determine that there are 152,892 pairs of twin primes less than thirty million.

In 1959, the IBM Corporation delivered the first IBM 7090 computer. The 7090 computer used many of the same parts (transistors, memories, tape controllers, circuit cards, frames, cables, power generators, etc.) as the Stretch (IBM 7030) computer. It was lower in price than Stretch and offered an operating system. The IBM 7090 could solve less complex problems more cheaply than Stretch. In 1961, an IBM 7090 was used to compute the 19th Mersenne prime number, the first known prime to have more than 1000 digits; and to compute π to 100,265 decimal places.

1958 In July, an IBM 704 computer at the Paris Data Processing Center was programmed according to a combination of Machin's formula and the Gregory series, to produce π to 10,000 decimal places in 1 hour and 40 minutes.

1959 American mathematician D.H. Lehmer showed that the number of primes below 10^{10} is 455,052,511.

1959 Francois Genuys programmed an IBM 704 computer at the Commissariat a l'Energie Atomique in Paris, to generate π to 16,167 decimal digits in 4.3 hours.

1960 The largest number known to be prime was $2^{3217} - 1$.

1961 An IBM 7090 computer at the London Data Centre was used to compute π to 20,000 decimal places in 39 minutes.

1961 The IBM 7090 computer can, in one second, perform 229,000 additions or subtractions, or 39,500 multiplications, or 32,700 divisions. The IBM 7090 is being used to perform many number theory calculations.

1961 American mathematician Alexander Hurwitz used an IBM 7090 computer to compute the 19th Mersenne prime, $2^{4253} - 1$, the first known prime to have more than 1000 digits. Hurwitz also found the 20th Mersenne prime, $2^{4423} - 1$.

1961 An IBM 7090 computer was used to determine two new perfect numbers (for $n = 4,253$ and 4,423 in Euclid's formula). There are no other even perfect numbers for $n < 5000$.

1961 Daniel Shanks and John Wrench used an IBM 7090 computer at the IBM Data Processing Center to compute π to 100,265 decimal places. The computations took 8 hours and 43 minutes. Shanks and Wrench used an arctangent formula by Carl Gauss to perform the calculation.

1962 A perfect magic cube of order 7 appeared in *Play Mathematics* by Harry Langman.

1963 D.B. Gillies discovered the 21st Mersenne prime, $2^{9689} - 1$, and the 22nd Mersenne prime, $2^{9941} - 1$.

1963 Selfridge discovers that the smallest known Sierpinski number is 78,557.

$$2^{11213} - 1$$

IS PRIME

A line of advanced computers were conceived and developed at the University of Illinois. ILLIAC I, a vacuum tube machine completed in 1952, could perform 11,000 arithmetical operations per second. ILLIAC II, a transistor-and-diode computer completed in 1963, could perform 500,000 operations per second. In 1963, the ILLIAC II computer was used to compute the 23rd Mersenne prime number with 3376 digits. The university personnel were so proud of this discovery that advertised it on the university's postage meter. This number remained the largest known prime for nine years.

1963 $2^{11213} - 1$ was proven prime at the University of Illinois. The University commemorated the event with a special postmark. This number remained the largest known prime for nine years. This 23rd Mersenne prime was discovered by D.B. Gillies.

1963   Selfridge and Hurwitz, by means of Pepin's test, showed that Fermat number F_{14} is composite, without finding any of its prime factors.

1963   Twenty-three perfect numbers are now known. The largest of these is $2^{11,212}(2^{11,213} -1)$, which contains 6,751 digits.

1964   E. Lehmer showed that there exists infinitely many Fibonacci pseudoprimes.

1964   Frenchman Paul Levy postulated that every odd number greater than 3 can be expressed in the form $2P + Q$ where P and Q are prime, for example, $27 = 2(11) + 5$.

1964   J.A.H. Hunter found a pair of 17-digit automorphic numbers. An automorphic number is one whose square ends with the given number: $5^2 = 25$; $76^2 = 5,776$; $625^2 = 390,625$. Thus 5, 76 and 625 are automorphic numbers.

1965   Stein and Stein verified Christian Goldbach's conjecture up to 10^8.

1966   In February, π was computed to 250,000 decimal places by French mathematicians Jean Gilloud and J. Fillatoire on an IBM 7030 computer (Stretch) at the Commissariat a l'Energie Atomique in Paris.

1966   Martin Gardner's mystical Dr. Matrix predicted that the millionth digit of π would be 5. Seven years later, Jean Gilloud and associates in Paris calculated π to one million decimal digits, using a Control Data CDC 7600 computer. Surprisingly, the millionth digit of π turned out to be 5.

1966   The best mathematicians can have incorrect intuition. For example, in 1753, Leonhard Euler thought that the exponent-three case of Fermat's Last Theorem, which he was first to prove, could be generalized to: If $n \geq 3$, then fewer than n nth powers cannot sum to an nth power. About two hundred years later, L.J. Lander and T.R. Parkin discovered by a computer search that $27^5 + 84^5 + 110^5 + 133^5 = 144^5$. And recently N. Elkies has shown that there are infinitely

In 1967, a Control Data Corporation's CDC 6600 computer was used by two French mathematicians to compute π to 500,000 decimal places.

many relatively prime triplets of fourth powers that sum to a fourth power. For example,

$$95800^4 + 217519^4 + 414560^4 = 422481^4$$

1967   Bryant Tuckerman (IBM Corporation) showed that any odd perfect number must be greater than 10^{36}.

1967   Lander and Parkin discovered the longest known (at this time) string of consecutive primes in arithmetic progression. It contained six terms, its difference is 30 and the initial term is 121,174,811.

1967   In February, a Control Data CDC 6600 computer was programmed by French mathematicians Jean Gilloud and Michele Dichampt to compute π to 500,000 decimal places in 28 hours and 10 minutes. The half-million values of π were published in a report from the Commissariat a l'Energie Atomique in Paris.

1969   In 1918, Poulet found a chain of five sociable numbers. This chain and the 28-link chain starting 14316 were the only known sociable chains until 1969 when, using computers, Henri Cohen checked all possible values for the smaller of the pair below 60 million and discovered 7 new chains, each of 4 links.

1970   The theory of odd perfect numbers is less comprehensive than for even ones, but there exists several results about the form of an odd perfect number must have, if one exists. If N is an odd perfect number, then D. Suryanarayana and P. Hagis found that the sum of the reciprocals of the prime divisors of N lies between 0.596 and 0.694.

1970   The solution to David Hilbert's tenth problem, a theorem that "is a remarkabale achievement of 20th century mathematicians and deserves the close attention of anyone interested in the fundamental nature of the mathematical enterprise," was completed when Yuri Matijasevic of the Soviet Union put the finishing touches on a body of work carried out by the American logicians Martin Davis, Hilary Putnam, and Julia Robinson.

1970   Morrison and Brillhart found the two prime factors of Fermat number F_7, which has 39 digits. The factors are
$F_7 = (2^9 \times 116{,}503{,}103{,}764{,}643 + 1)(2^9 \times 11{,}141{,}971{,}095{,}088{,}142{,}685 + 1)$.

Perfect Number	Number of Digits
1. $2^1(2^2 - 1) = 6$	1
2. $2^2(2^3 - 1) = 28$	2
3. $2^4(2^5 - 1) = 496$	3
4. $2^6(2^7 - 1) = 8,128$	4
5. $2^{12}(2^{13} - 1) = 33,550,336$	8
6. $2^{16}(2^{17} - 1) = 8,589,869,056$	10
7. $2^{18}(2^{19} - 1) = 137,438,691,328$	12
8. $2^{30}(2^{31} - 1)$	19
9. $2^{60}(2^{61} - 1)$	37
10. $2^{88}(2^{89} - 1)$	54
11. $2^{106}(2^{107} - 1)$	65
12. $2^{126}(2^{127} - 1)$	77
13. $2^{520}(2^{521} - 1)$	314
14. $2^{606}(2^{607} - 1)$	366
15. $2^{1278}(2^{1279} - 1)$	770
16. $2^{2202}(2^{2203} - 1)$	1,327
17. $2^{2280}(2^{2281} - 1)$	1,373
18. $2^{3216}(2^{3217} - 1)$	1,937
19. $2^{4252}(2^{4253} - 1)$	2,561
20. $2^{4422}(2^{4423} - 1)$	2,663
21. $2^{9688}(2^{9689} - 1)$	5,834
22. $2^{9940}(2^{9941} - 1)$	5,985
23. $2^{11212}(2^{11213} - 1)$	6,751
24. $2^{19937}(2^{19938} - 1)$	12,003
25. $2^{21700}(2^{21701} - 1)$	13,066
26. $2^{23208}(2^{23209} - 1)$	13,973
27. $2^{44496}(2^{44497} - 1)$	26,790
28. $2^{86242}(2^{86243} - 1)$	51,924
29. $2^{110502}(2^{110503} - 1)$	—
30. $2^{132048}(2^{132049} - 1)$	79,502
31. $2^{216090}(2^{216091} - 1)$	130,100
32. $2^{756838}(2^{756839} - 1)$	455,663
33. $2^{859432}(2^{859433} - 1)$	517,430

A number is said to be perfect if it is the sum of the divisors other than itself. Thus 6 and 28 are perfect numbers since 6 = 3 + 2 + 1, and 28 = 14 + 7 + 4 + 2 + 1. The first four perfect numbers are 6, 28, 496 and 8128. These four perfect numbers were known to Swiss mathematician Leonhard Euler in the 18th century. By 1971, twenty-four perfect numbers were known. The above table lists the known thirty-three perfect numbers.

1971   The twenty-fourth Mersenne prime has 6002 digits. The discovery of this number was reported by Bryant Tuckerman in the June issue of the *Notices* of the American Mathematical Society. An IBM 360/91 computer established primality of this large number in 39.44 minutes.

1971   David Singmaster found that 3003 is the smallest number to appear eight times in Pascal's triangle. There is no other number appearing so often, less than 2^{23}.

1971   Twenty-four perfect numbers are known.

1973   *A Handbook of Integer Sequences* was published by Academic Press. This book, compiled by N.J.A. Sloane of Bell Laboratories, lists more than 2300 integer sequences in numerical order.

1973   Richard Schroeppel, a mathematician and computer programmer, using a Digital Equipment Corp. PDP-11 minicomputer, determined that there are 275,305,224 magic squares of order 5 (excluding rotations and reflections).

1975   Whitfield Diffie and Martin Hellman invented the trapdoor function which made the factoring of very large numbers a matter of public interest and military concern. Shortly afterwards, Rivest, Shamir and Adleman showed how to make it a practical proposition. This is a mathematical function that will change any number A into its code number B. The function also has an inverse, which can be used to calculate A from B. The heart of these functions is a number that is the product of two large primes.

1975   Swift compiled a table of Carmichael numbers up to 10^9.

1975   The theory of odd perfect numbers is less comprehensive than for even ones, but there exists several results about the form of an odd perfect number must have, if one exists. If N is an odd perfect number, then P. Hagis found that N has at least 8 distinct prime factors; if N is not divisible by 3 then N has at least 11 prime factors.

1976   Martin Gardner (*Scientific American*, January) published an example of a magic cube constructed by Richard Myers. Myers discovered how to construct vast numbers of magic cubes by superimposing three Latin cubes and using octal notation when he was sixteen years old.

Shown here is a Control Data Corporation CDC 7600 computer. This computer was used by French mathematicians Jean Gilloud and Martine Bouyer to compute π to a million decimal places. The calculataion took 23 hours and 18 minutes.

1976   William Klein, the "human computer," before an audience at the CERN (European Center of Nuclear Research) auditorium, computed the 73rd root of a 499-digit number. At the end of the 2 minutes and 43 seconds he announced his answer, 6,789,235. His answer was confirmed by a computer. Since modern computers have overtaken Klein in speed, the great mental computer retired.

1976   The first supercomputer was designed by Seymour Cray at Cray Research. The Cray 1 supercomputer contains 200,000 integrated circuits and does 150,000,000 calculations per second.

1976   For many years, one of the most celebrated unsolved problems in mathematics has been the famous conjecture that four colors suffice to color any map on a plane or a sphere, where in the map no two countries sharing a common linear boundary can have the same color. An enormous amount of effort has been expended on this problem. In the summer of this year, Kenneth Appel and Wolfgang Haken of the University of Illinois, establish the conjecture by an immensely intricate computer-based analysis. The proof contains several hundred pages of complex detail and subsums and required over 1000 hours of computer calculations. Their proof was controversial in that a computer was necessary to check out a large number of special cases.

1976   Jean Gilloud and Martine Bouyer computed π to a million decimals, a calculation that took 23 hours and 18 minutes on a CDC 7600 computer. The results were published in a 400-page book.

1976   Richard Schroeppel and Ernst Straus independently found order-7 magic cubes.

1976   S.S. Wagstaff found that Fermat's Last Theorem is true for $n \leq 125,000$.

1977   Ronald Rivest of M.I.T., one of the three co-founders of RSA encryption, predicted that it would take more than 40 quadrillion years to factor RSA-129, a 129-digit number. In 1994, however, a team assisted by over 600 computers throughout the world succeeded in factoring this large number into a 64-digit number times a 65-digit number.

Several of the most secure cipher systems are based on the fact that large numbers are extremely difficult to factor, even using the most powerful computers for long periods of time. One cryptographic system based on the difficulty of factoring large numbers was invented in 1977 by Ronald L. Rivest, Adi Shamir and Leonard Adelman, all of the Massachusetts Institute of Technology. In the RSA System (which takes its initials from the names of its inventors), digits replace each letter in a text message, and the entire sequence of digits is treated as a single large number. A mathematical operation is then performed on this number, and to decipher the result requires either that the receiver possesses the key or breaks the code by factoring the large number.

1977 Hugh C. Williams together with Eric Seah discovered a prime number. It has 317 decimal digits, and it is a repunti number — each digit is a 1. It is the first new repunit prime discovered in 50 years.

1977 Weintraub discovered the longest known (at this time) string of consecutive primes in arithmetic progression; its difference is 30 and it begins with 999,900,067,719,989.

1977 To the Chinese this was the year of the snake. The Chinese have a 12-year cycle in which each year is named after an animal. They believe that people born in a particular year possess the characteristics of the animal associated with that year.

1977 Microcomputers from Apple Computer, Commodore, and Tandy Corporation (Radio Shack) started a boom in personal computers. Mathematicians, teachers, and students throughout the world started using these calculating machines to create many new records in number theory.

1977 Ivan Niven presented a talk at a number theory conference in which he mentioned a question concerning integers which are twice the sum of their digits. Since then, the properties of numbers which are divisible by their digital sum are called "Niven numbers."

1978 Two high school students, Curt Noll and Laura Nickel, using a Control Data Cyber 174 computer, discovered the 25th Mersenne prime. The discovery was national news — Walter Cronkite even read the story on the CBS Evening News. This large prime number had 6,533 digits.

1978 M. Yorinaga found eight Carmichael numbers with 13 prime factors — these are the ones with the largest number of prime factors known up to now.

1978 H.C. Williams found the prime repunit number R_{317}.

1978 The theory of odd perfect numbers is less comprehensive than for even ones, but there exists several results about the form of an odd perfect number must have, if one exists. If N is an odd perfect number then J. Condict found that N has a prime factor larger than 300,000.

1979 In February, Curt Noll used a Cyber 174 computer to discover M_{23209}, the 26th Mersenne prime which contained 6987 digits. It took Noll more than 8 hours to check this number on the computer. Two weeks

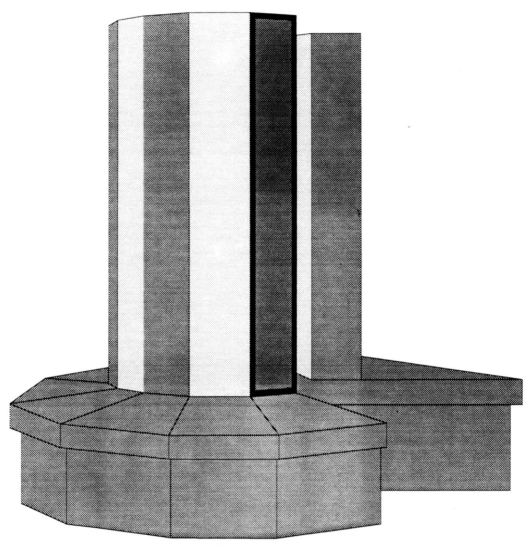

The first supercomputer (Cray 1) was designed by Seymour Cray at Cray Research, Inc. The Cray 1 has been used to produce many record breaking events in number theory. In 1979, David Slowinski and Harry Nelson used a Cray 1 supercomputer to discover the 27th Mersenne prime number with 13,395 digits. In 1982, Slowinski used a Cray 1 to find the 28th Mersenne prime with 28,962 digits. A year later, Slowinski used two Cray 1 supercomputers linked together to locate the 30th Mersenne prime with 39,751 digits.

later, David Slowinski used a Cray 1 supercomputer to check Noll's result, and the calculation was performed in 7 minutes. In April, David Slowinski and Harry Nelson, using a Cray 1, discovered M_{44497}, the 27th Mersenne prime with 13,395 digits.

1979   Cormack and Williams found the smallest known titanic prime, $25 \times 2^{3314} - 1$, with exactly 1,000 digits.

1979   Aiken and Rickert discovered the largest pair of twin primes: $1,159,142,985 \times 2^{2304} \pm 1$.

1979   M. Penk found a 14-digit factor of $2^{257} - 1$ (Marin Mersenne thought this number to be prime). What remained was still a 63-digit composite factor. Robert Baillie of the University of Illinois found this factor using a Control Data CDC 6600 computer.

1979   Hans Eberstark, an interpreter for the United Nations Agency, recited, at the European Nuclear Research Center, the decimal expansion of π to a total of 9744 places, breaking his previous record of 5050 places. Eberstark's rival in the memorization of π is David Sanker of the United States. The two vie with one another for listing in the *Guinness Book of World Records*.

1980   Adleman and Rumely developed a test that would decide if a randomly chosen number of up to 100 digits was prime in 4-12 hours with a large computer. This has been improved by Cohen and Lenstra to run about 1000 times faster. It can now test a 100-digit number in about 40 seconds on a supercomputer.

1980   Light, Forrest, Hammond and Roe, unaware of Stein and Stein's 1965 calculations, verified Goldbach's conjecture up to 10^8.

1980   Keller showed that Fermat number F_{9448} is composite, having the factor $19 \times 2^{9450} + 1$.

1980   In the November issue of *Crux mathematicorum* appears a table of all 93 five-digit and all 668 seven-digit palindromic primes (a palindromic number is a number that reads the same in both directions, as the palindromic prime 3417143). The calculation was done on a Digital Equipment Corporation PDP-11/45 minicomputer at the University of Waterloo, and the computer time was slightly more than one minute.

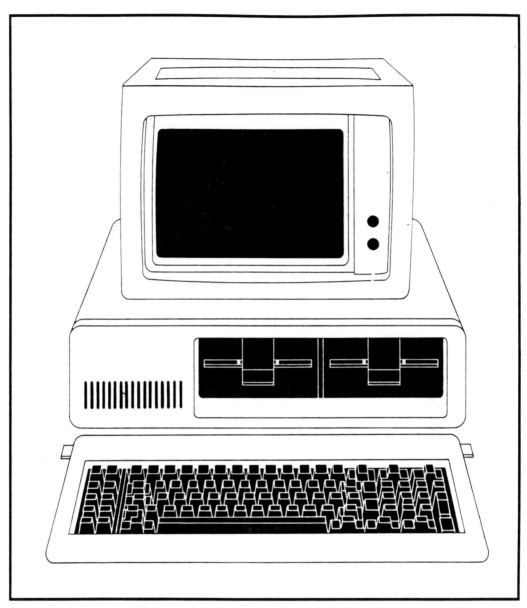

August 12, 1981, came and went, but nothing would ever be the same again. That day, the IBM Corporation introduced a Personal Computer based on the Intel 8088 microprocessor. The machine went on to become the most significant technology to hit the world since the telephone. It has transformed the way millions of people work, spawned new industries and made computer technology less mysterious. Personal computer users around the world are using IBM compatible personal computers to produce new number theory results on a regular basis. In 1985, a high school student used an IBM PC to demonstrate that there are no odd perfect numbers below 10^{79} that have eight prime divisors.

1981   Two Japanese mathematicians Kazunori Miyoshi and Kazuhika Nakayama of the University of Tsukuba calculated π to 2,000,038 significant figures in 137.30 hours on a FACOM M-200 computer.

1981   Personal computers became more powerful, easier to use and available at a reasonable price.

1981   R.P. Brent and John Pollard found the prime factor 1,238,926,361,552,897 of the Fermat number F_8.

1981   Rajan Mahadevan memorized π correctly to 31,811 decimal digits.

1981   The theory of odd perfect numbers is less comprehensive than for even ones, but there exists several results about the form of an odd perfect number must have, if one exists. If N is an odd perfect number, then P. Hagis found that the second largest prime factor of N is at least 1,000.

1982   A 25,962-digit number, the 28th Mersenne prime M_{86243}, was found by David Slowinski on a Cray 1 supercomputer. It took the Cray 1 a little over an hour to check this prime.

1982   The origin of Smith numbers appeared in a paper by Albert Wilansky that appeared in the January issue of the *Two-Year College Mathematics Journal*. A Smith number is a composite number the sum of whose digits is the sum of all the digits of all its prime factors: 4,937,775 = 3 x 5 x 5 x 65,837 and the digits in each expression sum to 42. It is not known if the number of Smith numbers is finite or infinite.

1982   Yoshiaki Tamura and Yasumasa Kanada of the University of Tokyo, using the HITAC M-280H supercomputer, calculated π to 4,194,293 decimal places. This computation required 2 hours and 53 minutes of computer time.

1982   D. Woods and J. Huenemann found a 432-digit Carmichael number.

1983   Yates coined the expression "titanic prime" to name any prime with at least 1000 digits. The following year Yates compiled a list of the 110 largest known titanic primes. By January 1985, he knew 581 titanic primes, of which 170 have more than 2000 digits. In September 1988, Yate's list contained 876 titanic primes. By 1990, another 550 primes were added to the list.

Shown here is a Cray X-MP supercomputer. In 1983, this computer was used by research scientists at Sandia Laboratories to factor several large numbers. Two years later, computer scientist David Slowinski used a Cray X-MP to find the 31st Mersenne prime number with 65,050 digits.

1983 A Japanese team of Yoshiaki Tamura and Tasumasa Kanada produced π to 8,388,608 places in 6.75 hours. Later in the year they computed π to 16,777,216 decimal digits in less than 30 hours.

1983 In September, David Slowinski using two Cray 1 supercomputers linked together discovered the 30th Mersenne prime, $2^{132049} - 1$, a 39,751 digit number.

1983 A group at Sandia Laboratories used a Cray X-MP supercomputer to factor several large numbers. Among them was $(10^{71} - 1)/9$ which, at the time, set a record for the largest number ever factored as the product of nearly equal primes.

1983 G. Faltings found that for every $n \geq 3$, Fermat's equation (Fermat's Last Theorem) has only finitely many primitive solutions.

1984 Keller found that the largest known prime of the form $k^2 \times 2^n + 1$ was $17^2 \times 2^{18502} + 1 = (17 \times 2^{9251})^2 + 1$. This is also the largest known prime of the form $n^2 + 1$.

1984 Harvey Dubner discovered a 1,281-digit prime number with all its digits equal to 0 or to 1.

1984 The First International Conference on Fibonacci Numbers and their applications was held at the University of Patras, Greece. Mathematicians from thirteen countries gathered to exchange information and ideas on mathematical topics relating to Fibonacci numbers.

1984 Keller showed that Cullen numbers C4713, C5795, C6611 and C18496 are prime.

1984 In his work of 1644, French number theorist Marin Mersenne stated, without proof, that Mersenne number M_{251} is composite. It was not until the nineteenth century that mathematicians finally proved Mersenne correct, by finding that M_{251} contains both 503 and 54,217 as prime factors. However, complete prime factorization of M_{251} occurred in 1984 when two researchers used a Cray supercomputer to locate all the factors.

1985 Computer scientist David Slowinski discovered the 31st Mersenne prime, $2^{216091} - 1$, during Labor Day weekend while testing a newly

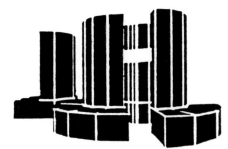

David Slowinski, a computer scientist at Cray Research Corporation, has used Cray supercomputers for generating several large Mersenne primes.

$(10^{103} + 1)/11 =$

$(1237) \times (44092859) \times (102860539) \times$

$(984385009) \times (612053256358933) \times$

$(182725114866521155647161) \times$

$(1471865453993855302660887614137521979)$

In 1984, A.L. Atkin and N.W. Rickert factored $(10^{103} + 1)/11$.

installed Cray X-MP 24 supercomputer at the Chevron Geosciences facility in Houston, Texas. This prime number contains 65,050 digits and took the supercomputer 3 hours to test that it was indeed a prime, following months of work to determine that it was a plausible candidate.

1985 Williams found a 3,021-digit prime number with one digit 1 and all other digits equal to 9, $2 \times 10^{3020} - 1$.

1985 Keller announced the largest known composite Fermat number is F_{23471}, which has the factor $5 \times 2^{23473} + 1$ and more than 10^{7000} digits.

1985 Carl Sagan in his novel *Contact* presents speculation about a hidden pattern or message God may have provided in the digits of π. In the story a supercomputer makes a discovery after countless hours of number crunching: the sequence of digits of π.

1985 Bill Gosper at M.I.T. computed π to 17 million decimal places.

1985 Lagarias, Miller and Odlyzko determined the largest known computed value of $\pi(x)$:
$$\pi(4 \times 10^{16}) = 1,075,292,778,753,150$$

1985 Michael Friedman used an IBM Personal Computer to demonstrate that there are no odd perfect numbers below 10^{79} that have eight prime divisors (which is the minimum number of prime divisors an odd perfect number could have).

1985 Harvey Dubner found a 1,057-digit Carmichael number.

1986 A Smith number with 2,592,699 digits was found.

1986 H. Williams and Harvey Dubner proved that the repunit consisting of 1,031 1s is a prime number.

1986 Keller determines the Sophie Germain prime, $q = 39051 \times 2^{6001} - 1$, and the composite Mersenne number Mq. To date, this is the largest known Sophie Germain prime and the largest known composite Mersenne number.

1986 In January, D.H. Bailey of the NASA Ames Research Center in California ran a Cray 2 supercomputer for 28 hours to obtain π to 29,360,000 decimal digits. His code was based on an algorithm by J.M. and P.D. Borwein of Dalhousie University.

I think it's another set of Smith numbers!

The Smith brothers — two consecutive Smith numbers. Smith numbers are composite numbers in which the sum of its digits is equal to the sum of all the digits of all its prime factors. An example of a Smith number is

$$9985 = 5 \times 1997$$
$$9 + 9 + 8 + 5 = 5 + 1 + 9 + 9 + 7$$

There are 47 of these numbers between 0 and 999, 32 between 1000 and 1999, 42 between 2000 and 2999, 28 between 3000 and 3999, 33 between 4000 and 4999, 32 between 5000 and 5999, 32 between 6000 and 6999, 37 between 8000 and 8999, and 40 between 9000 and 9999. The question of whether or not there are infinitely many Smith numbers remains open.

1986   In March, Harvey Dubner discovered that the largest known prime of the form $k \times 10^n + 1$ is $150{,}093 \times 10^{8000} + 1$. This prime number has 8,006 digits and it took approximately 60 days of computing time to find it.

1986   On July 3, Yasumasa Kanada of the University of Tokyo computed π to 33 million digits using 5 hours and 36 minutes of computer time.

1986   1,000,000! is known to have 5,565,709 decimal digits. It was found by Harry Nelson and David Slowinski.

1987   Jeff Young and Duncan Buell, using a Cray 2 supercomputer, proved that Fermat number F_{22} is composite.

1987   Young and Fry discovered that the longest known string of primes in arithmetic progression contains 20 primes, of which the smallest is 214,861,583,621 and the difference is 1,884,649,770.

1987   728 and 729 were found to be *Smith brothers*, consecutive Smith numbers. The palindromic Smith number, 12345554321 was also found.

1987   Yasumasa Kanada of the University of Tokyo using a NEC SX-2 supercomputer and an algorithm by J.M. and P.D. Borwein, calculated π to 134,217,728 decimal digits in 36 hours.

1987   Michael Kieth introduced the concept of replicating Fibonacci digits (called repfigits). At this time the largest repfigit discovered was the 7-digit number 7,913,837.

1987   A Japanese memorist recited 40,000 digits of π.

1987   The theory of odd perfect numbers is less comprehensive than for even ones, but there exists several results about the form of an odd perfect number must have, if one exists. If N is an odd perfect number, then G.L. Cohen found that the largest prime power dividing N is greater than 1020.

1987   J. Tanner and S.S. Wagstaff found that Fermat's Last Theorem is true for $n \leq 150{,}000$.

$$7532 \times \frac{10^{1104} - 1}{10^4 - 1} + 1$$

In 1988, Harvey Dubner discovered that the largest prime all of whose digits are prime numbers has 1104 digits.

$$\frac{11^{104} + 1}{11^8 + 1} =$$

(8675922231342839081221807709850708048977) x

(10848810485363747061296139984

297294840983461152579905772116753)

In 1988, A.K. Lenstra and M.S. Manasse factored a 100-digit number by a general purpose factorization algorithm.

1988   Young and Buell used a supercomputer to determine that Fermat number F_{20} is composite.

1988   Walter N. Colquitt and Luther Welsh, Jr., with the help of the NEC SX-2 supercomputer at the Houston Area Research Center, found the 29th Mersenne prime, M_{110503}. This prime occurs in a gap between two previous known Mersenne primes.

1988   Harvey Dubner discovered a 4,333-digit prime number with one digit 5 and all other digits equal to 9; $6 \times 10^{4333} - 1$. Dubner also found a 3,825-digit prime all of whose digits are odd, $1358 \times 10^{3821} - 1$.

1988   Yasumasa Kanada using an Hitachi supercomputer, calculated π to 201,326,000 decimal places in 6 hours.

1988   Harvey Dubner discovered a 1,104-digit prime all of whose digits are prime numbers.

1988   In October, two mathematicians succeeded in factoring a 100-digit number into two large prime factors, 41- and 60-digits long. The number factored was $11^{104} + 1$, and the method used was the quadratic sieve. The mathematicians were A.K. Lenstra and M.S. Manasse.

1988   In November, a 95-digit number was factored at the University of Georgia.

1988   The theory of odd perfect numbers is less comprehensive than for even ones, but there exists several results about the form of an odd perfect number must have, if one exists. If N is an odd perfect number, then R.P. Brent and G.L. Cohen found that $N > 10^{160}$.

1989   The Cray X-MP supercomputer is capable of 300 million calculations per second.

1989   Loh discovered that the largest known chain of Sophie Germain primes has length 12 and smallest prime 554,688,278,429.

1989   In June, two Columbia University research scientist brothers, David and Gregory Chudnovsky, computed π to 480 million decimal places.

1989   In August, a research team from Amdahl Corporation found a prime number that contained 65,087 decimal digits. The team consisted of

```
168 prime numbers between 1 and 1000
135 prime numbers between 1000 and 2000
127 prime numbers between 2000 and 3000
120 prime numbers between 3000 and 4000
119 prime numbers between 4000 and 5000
114 prime numbers between 5000 and 6000
117 prime numbers between 6000 and 7000
107 prime numbers between 7000 and 8000
110 prime numbers between 8000 and 9000
112 prime numbers between 9000 and 10000
106 prime numbers between 10000 and 11000
103 prime numbers between 11000 and 12000
109 prime numbers between 12000 and 13000
105 prime numbers between 13000 and 14000
102 prime numbers between 14000 and 15000
108 prime numbers between 15000 and 16000
98 prime numbers between 16000 and 17000
104 prime numbers between 17000 and 18000
94 prime numbers between 18000 and 19000
104 prime numbers between 19000 and 20000
98 prime numbers between 20000 and 21000
104 prime numbers between 21000 and 22000
100 prime numbers between 22000 and 23000
104 prime numbers between 23000 and 24000
94 prime numbers between 24000 and 25000
98 prime numbers between 25000 and 26000
101 prime numbers between 26000 and 27000
94 prime numbers between 27000 and 28000
98 prime numbers between 28000 and 29000
92 prime numbers between 29000 and 30000
95 prime numbers between 30000 and 31000
92 prime numbers between 31000 and 32000
106 prime numbers between 32000 and 33000
100 prime numbers between 33000 and 34000
94 prime numbers between 34000 and 35000
92 prime numbers between 35000 and 36000
99 prime numbers between 36000 and 37000
94 prime numbers between 37000 and 38000
90 prime numbers between 38000 and 39000
96 prime numbers between 39000 and 40000
88 prime numbers between 40000 and 41000
101 prime numbers between 41000 and 42000
102 prime numbers between 42000 and 43000
85 prime numbers between 43000 and 44000
96 prime numbers between 44000 and 45000
86 prime numbers between 45000 and 46000
90 prime numbers between 46000 and 47000
95 prime numbers between 47000 and 48000
89 prime numbers between 48000 and 49000
98 prime numbers between 49000 and 50000
```

Prime numbers become scarce when computed in large number ranges.

John Brown, Curt Noll, B.K. Parady, Gene Smith, Joel Smith, and Sergio Zarantonello. The number was found by multiplying 2 by itself 216,193 times, multiplying this by 391,581 and then subtracting 1. The new algorithm used for this high-speed convolution may also have applications in seismic research, weather prediction, aeronautical simulation, and the search for pulsars. An Amdahl mainframe computer was used to compute this large prime number.

1989 Harvey Dubner found the two largest known palindromic primes.

1989 Three new larger repfigits were discovered, the largest being 44,121,607.

1989 Yasumasa Kanada calculated π to 536,870,000 decimal places.

1989 In February, Y. Miyaoka (Tokyo Metropolitan) thought he had a proof of Fermat's Last Theorem.

1989 Two Columbia University research scientist brothers, David and Gregory Chudnovsky computed π to 1,011,196,691 decimal places. Printed out in a line, the digits they computed would stretch nearly halfway across the United States. The Chudnovsky's used Cray 2 and IBM 3090 mainframe computers to perform their computations.

1989 The twin primes, $1,706,595 \times 2^{11235} \pm 1$, with 3,389 digits each, were found by Brown, Noll, Parady, Smith, Smith, and Zarantonello.

1989 Young and Potler calculated the largest difference between consecutive primes to be 90,874,329,412,297.

1989 Clifford Pickover discovered two replicating Fibonacci digits (repfigits) in the range 100 million to 1 billion (129,572,008 and 251,133,297). These are believed to be the largest replicating Fibonacci digits discovered to date.

1989 Harvey Dubner found a 3,710-digit Carmichael number that is the product of the three primes with 929, 929 and 1,853 digits. The computation took 20 hours on a computer.

1989 Loh found that the largest known Cunningham chain of primes has length 13 and smallest prime 758,083,947,856,951.

CARMICHAEL NUMBERS

Carmichael numbers are the composite numbers n such that $a^{n-1} \equiv$ (mod n) for every integer $a, 1 < a < n$, such that a is relatively prime to n. Carmichael showed that if p is any prime dividing n then $p - 1$ divides $n - 1$. It follows that every Carmichael number is odd and the product of three or more distinct prime numbers. In 1939, Chernik gave the following method to obtain Carmichael numbers. Let $m \geq 1$ and $M_3(m)$ $= (6m + 1)(12m + 1)(18m + 1)$. If m is such that all three factors above are prime, then $M_3(m)$ is a Carmichael number. The smallest twenty Carmichael numbers are:

$561 = 3 \times 11 \times 17$ $41041 = 7 \times 11 \times 13 \times 41$

$1105 = 5 \times 13 \times 17$ $46657 = 13 \times 37 \times 97$

$1729 = 7 \times 13 \times 19$ $52633 = 7 \times 73 \times 103$

$2465 = 5 \times 17 \times 29$ $62745 = 3 \times 5 \times 47 \times 89$

$2821 = 7 \times 13 \times 31$ $63973 = 7 \times 13 \times 19 \times 37$

$6601 = 7 \times 23 \times 41$ $75361 = 11 \times 17 \times 31$

$8911 = 7 \times 19 \times 67$ $101101 = 7 \times 11 \times 13 \times 101$

$10585 = 5 \times 29 \times 73$ $115921 = 13 \times 37 \times 241$

$15841 = 7 \times 31 \times 73$ $126217 = 7 \times 13 \times 19 \times 73$

$29341 = 13 \times 37 \times 61$ $162401 = 17 \times 41 \times 233$

There are 43 Carmichael numbers less than one million. In 1989, Harvey Dubner found a 3710-digit Carmichael number that is the product of the three primes with 929, 929 and 1853 digits. The computation took 20 hours on a computer. In 1985, Dubner found a 1057-digit Carmichael number. In 1982, D. Woods and J. Huenemann found a 432-digit Carmichael number.

In 1912 R.D. Carmichael discovered a rare kind of number. In 1990 Jaeschke compiled a table of Carmichael numbers up to 10^{12}.

1989 The Japanese electronics firm NEC announced its SX-X supercomputer with an expected speed of 20 billion calculations per second.

1989 The prime number found this year by the Amdahl research team was the first time since 1952 that the largest known prime has not been a Mersenne prime. This new prime has 37 more digits than $2^{216091} - 1$, the largest known Mersenne prime.

1989 The theory of odd perfect numbers is less comprehensive than for even ones, but there exists several results about the form of an odd perfect number must have, if one exists. If N is an odd perfect number, then R.P. Brent, G.L. Cohen, and H.J.J. te Riele found that $N > 10^{300}$.

1990 H.J. te Riele used a single processor on a NEC-SX-2 to factor a 101-digit number. It was the first time that a difficult 100+ digit number was ever factored using just a single computer for the calculation.

1990 Jaeschke compiled a table of Carmichael numbers up to 10^{12}.

1990 Mathematicians develop a scheme for speeding up multiplication and a unique computer. "Little Fermat" works with 257-bit words and uses arithmetic based on Fermat numbers; it does arithmetic faster than the usual floating point algorithms while avoiding the errors of rounding. The machine, designed by David and Gregory Chudnovsky (Columbia University) and M.M. Denneau (IBM) was assembled by Saed G. Younnis. The machine is ideal for number-theoretic calculations but also for digital signal and image processing, as well as for solving differential equations.

1990 The 155-digit Fermat number $F_9 = 2^{512} + 1$ was factored by using the number field sieve invented by John Pollard. The factoring of F_9 took two months on 1,000 computers around the world. The ninth Fermat number was factored to 2,424,833 (a prime) and a 99- and 49-digit prime number. The method uses a new algorithm that is based on factorization of integers in algebraic number fields; the method applies to integers of the form $a^k + b$ for a and b very small and k very large. For the problem just solved, the results done in parallel were assembled to construct a sparse linear system with more than 200,000 equations, which was then reduced to a dense system of 72,000 equations by a structured Gaussian elimination.

Prime Numbers

3	5	7	11	13	17	19	23	29	31	37
41	43	47	53	59	61	67	71	73	79	83
89	97	101	103	107	109	113	127	131	137	139
149	151	157	163	167	173	179	181	191	193	197
199	211	223	227	229	233	239	241	251	257	263
269	271	277	281	283	293	307	311	313	317	331
337	347	349	353	359	367	373	379	383	389	397
401	409	419	421	431	433	439	443	449	457	461
463	467	479	487	491	499	503	509	521	523	541
547	557	563	569	571	577	587	593	599	601	607
613	617	619	631	641	643	647	653	659	661	673
677	683	691	701	709	719	727	733	739	743	751
757	761	769	773	787	797	809	811	821	823	827
829	839	853	857	859	863	877	881	883	887	907
911	919	929	937	941	947	953	967	971	977	983
991	997	1009	1013	1019	1021	1031	1033	1039	1049	1051
1061	1063	1069	1087	1091	1093	1097	1103	1109	1117	1123
1129	1151	1153	1163	1171	1181	1187	1193	1201	1213	1217
1223	1229	1231	1237	1249	1259	1277	1279	1283	1289	1291
1297	1301	1303	1307	1319	1321	1327	1361	1367	1373	1381
1399	1409	1423	1427	1429	1433	1439	1447	1451	1453	1459
1471	1481	1483	1487	1489	1493	1499	1511	1523	1531	1543
1549	1553	1559	1567	1571	1579	1583	1597	1601	1607	1609
1613	1619	1621	1627	1637	1657	1663	1667	1669	1693	1697
1699	1709	1721	1723	1733	1741	1747	1753	1759	1777	1783
1787	1789	1801	1811	1823	1831	1847	1861	1867	1871	1873
1877	1879	1889	1901	1907	1913	1931	1933	1949	1951	1973
1979	1987	1993	1997	1999	2003	2011	2017	2027	2029	2039
2053	2063	2069	2081	2083	2087	2089	2099	2111	2113	2129
2131	2137	2141	2143	2153	2161	2179	2203	2207	2213	2221
2237	2239	2243	2251	2267	2269	2273	2281	2287	2293	2297
2309	2311	2333	2339	2341	2347	2351	2357	2371	2377	2381
2383	2389	2393	2399	2411	2417	2423	2437	2441	2447	2459
2467	2473	2477	2503	2521	2531	2539	2543	2549	2551	2557
2579	2591	2593	2609	2617	2621	2633	2647	2657	2659	2663
2671	2677	2683	2687	2689	2693	2699	2707	2711	2713	2719
2729	2731	2741	2749	2753	2767	2777	2789	2791	2797	2801
2803	2819	2833	2837	2843	2851	2857	2861	2879	2887	2897
2903	2909	2917	2927	2939	2953	2957	2963	2969	2971	2999
3001	3011	3019	3023	3037	3041	3049	3061	3067	3079	3083
3089	3109	3119	3121	3137	3163	3167	3169	3181	3187	3191
3202	3209	3217	3221	3229	3251	3253	3257	3259	3271	3299
3301	3307	3313	3319	3323	3329	3331	3343	3347	3359	3361
3371	3373	3389	3391	3407	3413	3433	3449	3457	3461	3463
3467	3469	3491	3499	3511	3517	3527	3529	3533	3539	3541
3547	3557	3559	3571	3581	3583	3593	3607	3613	3617	3623

The first 506 odd prime numbers.

1990   B.K. Parody, Joel Smith, and Sergio Zarantonello determined a large pair of twin primes: $571,305 \times 2^{7701} \pm 1$.

1990   All amicable number pairs below one billion have now been found.

1990   Robert Dubner of Dubner Computer Systems published *The Book of Primes*, a book containing prime numbers to 10 million.

1991   J.P. Buhler, Candall, and R.W. Sompolski found that Fermat's Last Theorem is true for $n \leq 1,000,000$.

1991   Donald L. Miller and Joseph F. Pekny identify an exact algorithm for optimal solutions to the traveling salesman problem with *asymmetric* intercity costs, which finds solutions to random asymmetric problems with 5,000 cities in a few seconds and with 500,000 cities in a few hours. The algorithm reported is used to generate schedules for a variety of chemical manufacturing facilities, a problem for which asymmetric costs are typical.

1991   The Mersenne numbers, $M_p = 2^p - 1$, have been the object of much study for a long time. Earlier research had found the 28 Mersenne primes for M_p for $p < 100,000$. In addition, M_{132049} and M_{216091} were known to be prime. W.N. Colquitt and L. Welsh, Jr., with the help of the NEC SX-2 supercomputer at the Houston Area Research Center, undertook a systematic search for Mersenne primes M_p with p in the intervals $100,000 < p < 139,268$ and found that the only other Mersenne exponent between 100,000 and 139,268 is 132,049. Thus, M_{110503} and M_{132049} are the 29th and 30th Mersenne primes in order of size, respectively. Finally, it was shown that there are no Mersenne exponents between 216,092 and 353,620. (It is not known whether a Mersenne prime exists in the interval $139,268 < p < 216,090$.)

1991   Harvey Dubner found the largest known palindromic prime, $10^{11310} + 4,661,664 \times 10^{56752} + 1$, with 11,311 digits.

1991   The record-largest traveling problem that has been solved exactly is one with 2,392 cities: researchers are now competing to solve a 3,038-city problem. Traveling salesman problems of enormous size arise in the fabrication of circuit boards and very large-scale integrated circuits, in which as many as a million holes ("cities") need to be drilled. For a million-city tour, it now takes about 3.5 hours of computing to get an answer that is within 3.5% of optimal, and seven months to get within 0.75%.

	DECIMAL PLACES
Archimedes, 240 B.C.	2
Cladius Ptolemy, 150 A.D.	4
Tsu Ch'ung Chih, 480 A.D.	6
Leonardo Fibonacci, 1220	4
Valentinus Otho, 1573	7
Francois Viete, 1593	10
Adriaen van Roomen, 1593	15
Ludolph van Ceulen, 1610	35
Abraham Sharp, 1705	72
John Machin, 1706	100
DeLagny, 1717	112
Georg Vega, 1794	140
William Rutherford, 1824	153
von Strassnitsky and Dase, 1844	200
Thomas Clausen, 1847	248
William Rutherford, 1853	440
Richter, 1855	500
William Shanks, 18873	527
Ferguson, 1946	620
Ferguson, 1947	710
Ferguson, 1947	808
John Wrench and Levi Smith, 1949	1120
Reitwiesner and ENIAC, 1949	2037
Nicholsen, Jeenel and NORC, 1954	3089
Felton and PEGASUS, 1957	7480
Genuys, (IBM 704), 1958	10,000
Genuys, (IBM 704), 1959	16,167
(IBM 7090), 1961	20,000
Shanks and Wrench, (IBM 7090), 1961	100,265
Gilloud and Fillatoire, (IBM 7030), 1966	250,000
Gilloud and Dichampt, (CDC 6600), 1967	500,000
Gilloud and Bouyer, (CDC 7600), 1976	1,000,000
Miyoshi and Nakayama, (FACOM M-200), 1981	2,000,038
Tamura and Kanada, (HITAC M-280H), 1982	4,194,293
Tamura and Kanada, (HITAC M-280H), 1983	8,388,608
Tamura and Kanada, (HITAC M-280H), 1983	16,777,216
Gosper (1985)	17,000,000
Bailey, Borwein, Borwein, (Cray 2), 1986	29,360,000
Kanada, 1986	33,000,000
Kanada, (NEC SX-2), 1987	134,217,728
Kanada, (Hitachi supercomputer), 1988	201,326,000
Chudnovsky, David, and Gregory, (Cray 2 and IBM 3090), 1989	480,000,000
Kanada, (Hitachi supercomptuer), 1989	536,870,000
Chudnovsky, David and Gregory (Cray 2 and IBM 3090), 1989	1,011,196,691

Key milestones in the computation of π.

1992 David and Gregory Chudnovsky have designed and built a supercomputer in their apartment from mail-order parts and used it to generate and check for patterns in two billion digits of π. So far, no patterns. In 1989, these two Soviet brothers computed π to over a billion digits.

1992 Enrico Bombieri, one of the world's leading number theorists, shows how the distribution of primes can be analyzed by means of harmonic analysis of waveforms. Bombieri also goes on to state that the Riemann Hypothesis is the most important unsolved problem in mathematics today.

1992 In February, a new record prime was found by David Slowinski and Paul Gage using a Cray 2 supercomputer at the Harwell Laboratory of AEA Technology near Oxford, England. The prime, $2^{756839} - 1$, is now the largest known prime. It has 227,832 digits — over 3.5 times as many digits as the previous record found 1989. After 19 hours of searching, the Cray 2 found the 32nd known Mersenne prime. The computer program used, which was designed by David Slowinski, is based on the Lucas-Lehmer test. By the theorem in Euclid's *Elements* Book XIII, the new Mersenne prime also yields the 32nd known perfect number. The new perfect number has 455,663 digits.

1992 There are 43 Carmichael numbers less than one million. In 1910, Carmichael listed 15 of them and footnoted that "This list might be indefinitely extended." His use of "might" seems to imply uncertainty. Other mathematicians were uncertain, too, until now. In September, Red Alford, Andrew Granville, and Carl Pomerance (University of Georgia) proved that for sufficiently large N, there are more than $N^{2/7}$ Carmichael numbers.

1993 In July, newspapers and magazines reported that the most famous outstanding problem in mathematics had finally been solved. Andrew Wiles of Princeton University had proved Fermat's Last Theorem (FLT) which states that the equation $x^n + y^n = z^n$ has no solution in positive integers x, y, z, n for $n > 2$. Wiles, however, faced a troubling gap in logic he had followed to prove Fermat's Last Theorem.

1993 A multiperfect number is one whose divisors add up to a multiple of the number; the multiple is called the *index*. For example, 360 is multiperfect with index 3. Until recently, only 700 multiperfects were known. An algorithm of Fred Helenius in Colorado has almost doubled

FERMAT NUMBERS

F_0 = 3 (prime)
F_1 = 5 (prime)
F_2 = 17 (prime)
F_3 = 257 (prime)
F_4 = 65537 (Largest known Fermat prime)
F_5 = composite: 641 x 6700417
F_6 = composite: 274177 x 67280421310721
F_7 = composite
F_8 = composite
F_9 = composite: 2^{512} + 1
F_{10} = 455925777 x 6487031807 x composite number
F_{11} = composite: 2^{2948} + 1
F_{14} = composite, factors unknown
F_{20} = composite, factors unknown
F_{22} = prime or composite — ?
F_{24} = prime or composite — ?
F_{28} = prime or composite — ?
F_{37} = prime or composite — ?
F_{9448} = composite with the factor 19 x 2^{9450} + 1
F_{23471} = composite with the factor 5 x 2^{23473} + 1

Fermat numbers are numbers of the form $2^m + 1$. If $2^m + 1$ is a prime, then m must be of the form $m = 2^n$. Shown above are several Fermat numbers and an indication of whether they are prime or composite. The first five Fermat numbers are the only known Fermat primes.

the number of known multiperfects, including finding 14 with index 9. The algorithm starts with the product of powers of small primes and successively adjusts the number to find a multiperfect.

1994   Harry J. Smith wrote an IBM-compatible microcomputer program called *Perfect* that computes a perfect number when given the exponent of a Mersenne prime.

1994   Largest Fermat numbers and their factorization: $F_{11} = 2^{2048} + 1$ was factored by Brent and Morain in 1988. $F_9 = 2^{512} + 1$ was factored by Pollard in 1990. The factorization for F_{10} is still not known. F_{22} is the smallest Fermat number not yet proven prime or composite. In 1988, Young and Buell used a supercomputer to determine that F_{20} is composite. $F_4 = 65,537$ is the largest known Fermat prime.

1994   Harvey Dubner found the largest known twin primes $1,691,232 \times 1,001 \times 10^{4020} + 1$ and $1,691,232 \times 1,001 \times 10^{4020} - 1$. Dubner also found the twin primes $4,650,828 \times 1,001 \times 10^{3429} + 1$ and $4650828 \times 1001 \times 10^{3429} - 1$.

1994   The number, 114,381,625,757,888,867,669,235,779,976,146, 612,010,218,296,721,242,362,562,561,842,935,706,935,245, 733,897,830,597,123,563,958,705,058,989,075,147,599,290,026, 879,543,541, is better known as RSA-129, after the RSA cryptosystem and the number's 129 digits. The inventors of RSA proposed in 1977 that it would take others at least 40 quadrillion years to discover the factors of RSA-129. In May, however, a team assisted by 600 computers throughout the world succeeded in factoring this large number into a 64-digit number times a 65 digit number. This huge computation culminated eight months of work — a lot less time than originally predicted. The methods used are ordinarily reserved for numbers of 155 or more digits. However, this number was considered a particular challenge for reasons relating to encryption and security.

1994   On January 4, David Slowinski and Paul Gage of Cray Research announced a new largest Mersenne prime, $2^{859433} - 1$. It was found by a Cray Y-MP M90 series supercomputer using the traditional Lucas-Lehmer test after 7.2 hours. This 258,716 digit number is the 33rd Mersenne prime according to size. Other smaller ones may yet lie in the exponent gaps 386,000-430,000 and 524,000-750,000.

1994   In October, mathematician Andrew Wiles of Princeton University caught the mathematical community by surprise, when he began

Mersenne prime #	M_p where $p =$
1st	2
2nd	3
3rd	5
4th	7
5th	13
6th	17
7th	19
8th	31
9th	61
10th	89
11th	107
12th	127
13th	521
14th	607
15th	1279
16th	2203
17th	2281
18th	3217
19th	4253
20th	4423
21st	9689
22nd	9941
23rd	11213
24th	19937
25th	21701
26th	23209
27th	44497
28th	86243
29th	110503
30th	132049
31st	216091
32nd	756839
33rd	859433

French number theorist, Father Marin Mersenne (1588-1648) maintained a constant correspondence with the greatest mathematicians of his day and served admirably in those prejournal times, as a clearing house for mathematical ideas. He is especially known today in connection with Mersenne primes, or prime numbers of the form $2^p - 1$, which he discussed in several places in his work *Cogitata physico-mathematica* of 1644. The connection between Mersenne primes and perfect numbers was also pointed out in this work.

The known Mersenne primes.

distributing copies of two new manuscripts addressing the concerns that had been raised about his 1993 proof of Fermat's Last Theorem. The first paper announces the revised proof while the second paper, produced in collaboration with Cambridge mathematician Richard L. Taylor, contains mathematical reasoning justifying a key step in the main proof.

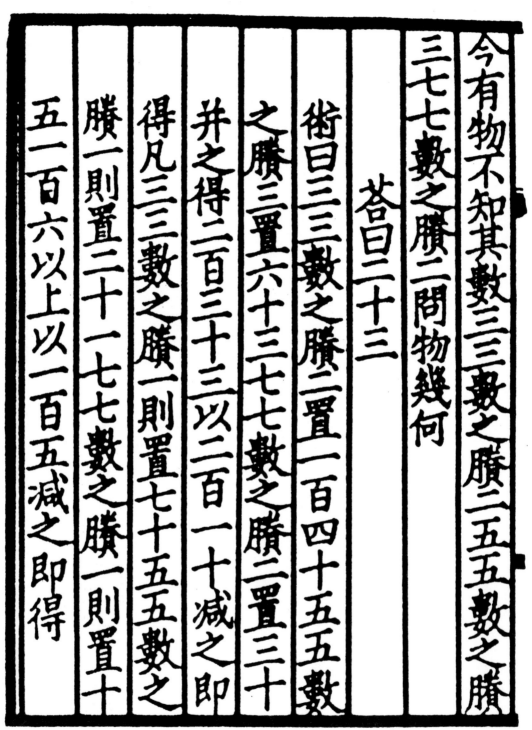

今有物不知其數三三數之賸二五五數之賸
三七七數之賸二問物幾何
荅曰二十三
術曰三三數之賸二置一百四十五五數
之賸三置六十三七七數之賸二置三十
并之得二百三十三以二百一十減之即
得凡三三數之賸一則置七十五五數之
賸一則置二十一七七數之賸一則置十
五一百六以上以一百五減之即得

The earliest known formulation of the Chinese remainder theorem appears
to be in the *Sun Tzu Suan-ching* (i.e., the mathematical classic of Sun Tzu)
which has been dated between 280 A.D. and 473 A.D.

Pascal's triangle appears on the title page of a 1527 arithmetic work by Petrus Apianus. In this work he introduced notation for multiplying and dividing.

Title page of a 1529 edition of an arithmetic book by Adam Riese. Riese was the most influential German writer in the move to replace the old computation (counters and Roman numerals) by the newer method (using pen and Hindu-Arabic numerals).

PICTURE CREDITS

All illustrations are from the picture archives of the Camelot Publishing Company except as noted below.

Control Data Corporation, 78, 82

Cray Research, Inc., page 90

IBM Corporation, pages 6, 8, 74

National Bureau of Standards, page 70

University of Illinois, page 76

U.S. Naval Weapons Laboratory, page 72

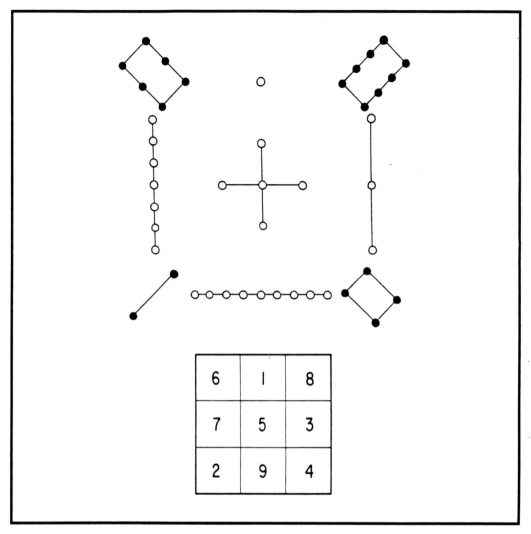

6	1	8
7	5	3
2	9	4

The first known example of a magic square comes from China. Legend tells us that around the year 2000 B.C. the Emperor Yu of the Shang dynasty received a magic square etched on the back of a tortoise's shell. This incident supposedly took place along the Lo River, so this magic square has come to be known as the Lo-shu magic square. Being the earliest and the simplest of all magic squares, this square has appeared more frequently than any other. The numerals in it later took the form of knots in strings, as shown above, in which black knots are used for the even numbers and white knots for the odd numbers. In the western world we find the first mention of magic squares in the work of Theon of Smyrna in 130 A.D. In the 9th century, magic squares crept into the world of astrology with Arab astrologers using them in horoscope calculations. Finally, with the works of the Greek mathematician Moschopoulos, in 1300 A.D., magic squares and their respective properties spread to the western hemisphere.

BIBLIOGRAPHY

This bibliography contains a number of books allied in interest to the contents of this book. It is not (nor is it intended to be) full, authoritative, or exhaustive. It is merely an expression of personal taste which may be helpful to the reader whose curiosity about number theory has been stimulated.

A History of Mathematics. Carl B. Boyer. Princeton University Press, 1985.

A History of π (Pi). Petr Beckmann. St. Martin's Press, 1971.

Ancient Puzzles. Dominic Olivastro. Bantam Books, 1993

Beyond Numeracy. John Allen Paulos. Alfred A. Knopf, Inc., 1991.

Computers In Number Theory. Donald D. Spencer. Computer Science Press, 1982

Discovering Number Theory. John E. and Margaret W. Maxfield. W.B. Saunders Company, 1972.

Elementary Number Theory: A Computer Approach. Allen M. Kirch. Intext Educational Publishers, 1974.

Elementary Number Theory. Underwood Dudley. W.H. Freeman and Company, 1969.

e: The Story of a Number. Eli Major. Princeton University Press, 1994.

Excursions In Number Theory. Stanley C. Ogilvy. Dover Publications, 1966.

Exploring Number Theory With Microcomputers. Donald D. Spencer. Camelot Publishing Company, 1991.

History of Mathematics, Vols. I and II. David E. Smith. Ginn & Company, Vol. I, 1923; Vol. II, 1925.

Introduction to the Theory of Numbers. Leonard Eugene Dickson. Dover Pubications, 1957.

Invitation To Number Theory With Pascal. Donald D. Spencer. Camelot Publishing Company, 1989.

Islands of Truth. Ivars Peterson. W.H. Freeman and Company, 1990.

Magic Cubes: New Recreations. William H. Benson and Oswold Jacoby. Dover Publications, 1981.

Magic Squares and Cubes. W.S. Andrews. Dover Publications, 1960.

Mathematics A Historical Development. Lee Emerson Boyer. Henry Holt and Company, 1945.

Mathematics and the Imagination. Edward Kasner and James R. Newman, Tempus Books, 1989.

Mathematics: The New Golden Age. Keith Devlin. Penguin Books, 1988.

New Recreations With Magic Squares. William H. Benson and Oswald Jacoby. Dover Publications, 1976.

Number. John McLeisk. Fawcett Columbine, 1991.

Numbers: Their History and Meaning. Graham Flegg. Penguin Books, 1983.

Number Theory And Its History. Oystein Ore. McGraw-Hill Book company, 1948.

Number Theory in Science and Communication. M.R. Schroeder. Springer-Verlag, 1984.

Number Treasury. Stanley J. and Margaret Kenney. Dale Seymour Publications, 1982.

Old and New Unsolved Problems In Plane Geometry and Number Theory. Victor Klee and Stan Wagon. Mathematical Association of America, 1991.

Pi in the Sky. John D. Barrow. Clarendon Press, 1992.

Recreations in the Theory of Numbers. Albert H. Beiler. Dover Publications, 1964.

The Little Book of Big Primes. Paulo Ribenboim. Springer-Verlag, 1991.

The Lore of Large Numbers. Philip J. Davis. Random House, 1961.

The Mathematical Experience. Philip J. Davis and Reubin Hersh. Houghton Mifflen Company, 1981.

The Mathematical Tourist. Ivars Peterson. W. H. Freeman and Company, 1988.

The Mathematical Traveler Exploring the Grand History of Numbers. Calvin C. Clawson. Plenum Press, 1994.

The Most Beautiful Mathematical Formulas. Lionel Salem, Frederic Testard, and Coralie Salem. John Wiley & Sons, 1992.

The Penguin Dictionary of Curious and Interesting Numbers. David Wells. Penguin Books, 1986.

The Wonders of Magic Squares. Jim Moran. Vintage Books, 1981.

Those Amazing Reciprocals. Boyd Henry. Dale Seymour Publications, 1992.

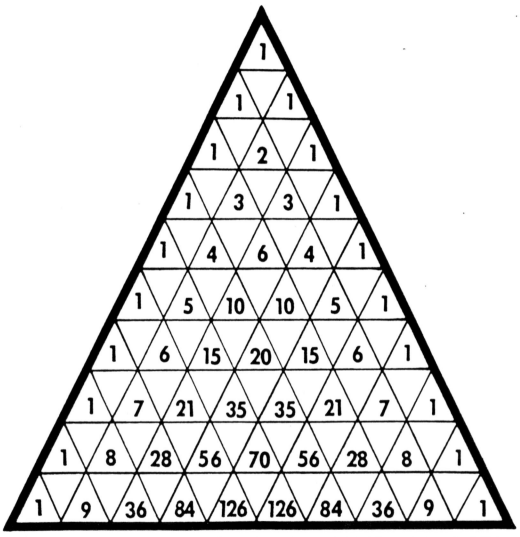

Pascal's triangle — a pattern of numbers from which binomial probabilities can be easily determined. It was known to Omar Khayyam about 1100 A.D. and was published in China about 1300 A.D. However, it is generally known as Pascal's triangle because of the amount of work he did on it.

INDEX

52	61	4	13	20	29	36	45
14	3	62	51	46	35	30	19
53	60	5	12	21	28	37	44
11	6	59	54	43	38	27	22
55	58	7	10	23	26	39	42
9	8	57	56	41	40	25	24
50	63	2	15	18	31	34	47
16	1	64	49	48	33	32	17

Benjamin Franklin was interested in magic squares. While in the Pennsylvania Assembly, he often amused himself with magic squares. He also thought that mathematical demonstrations are better than academic logic for training the mind to reason with exactness and distinguish truth from falsity, even outside mathematics. Shown above is a magic square concocted by Benjamin Franklin. He included it in a letter to Peter Collinson in about 1750. This 8 by 8 magic square, contains all the numbers from 1 to 64, arranged so that any row, column or major diagonals adds up to 260.